From Plough to Plane
My Autobiography

by

Peter Ellis Brown

Bound Biographies

On my way home from RAF service in Egypt, 1953

Copyright Peter Ellis Brown © 2009

Produced in association with

Bound Biographies

Heyford Park House, Heyford Park, Bicester, OX25 5HD

www.boundbiographies.com

ISBN: 978-1-905178-25-4

Dedication

This book is dedicated to my dear wife,
for always being there for me.

Introduction

Why did I write my life story? Over the years since both my dad and my eldest brother have no longer been here, I have always wanted to know more about their lives, especially about their military service days - Dad during the First World War, and my brother during the Second World War. I have their service records, but I know from mine that they only tell a small part of their story.

People over the years have often said to me, "With all the tales you have to tell, why don't you write them down?" So here they are, and I hope that my family and friends will enjoy the book.

In addition, I hope that future generations to come will also get some enjoyment from my book. I would like them to have some idea of how my generation lived and the times we lived in.

Contents

My Childhood in Suffolk	19
My Air Force Days Begin	45
214 Squadron in Kenya: 1954	63
Civvy Street and Aviation Traders	77
Carvair to Australia - Nearly	91
Engine Changes over the Years	99
My Years with the 'Sally B'	121
My Part in the Making of 'Memphis Belle'	147
HeavyLift Cargo Airlines	159
The Changing Village	183
The One Place Where It All Comes Together	193
Summing It All Up	197

Illustrations

Peter on his way home from RAF service in Egypt, 1953	3
Peter, age 5 (first school photo)	19
Market Weston School	20
The Market Weston salvage collectors	27
Barningham School, 1943	29
Map showing the bombing of Hopton, November, 1940	30
Map showing the bombing of Market Weston, 1941	31
Peter's two geese outside his home in Market Weston, 1947	36
Peter outside the house where he was born, Fen Street, Hopton	40
All the family with Mum, Market Weston Village Hall, 1980	40
Keeping my hand in on the tractor!	41
RAF Cardington, 1950 (first photograph taken in the RAF)	46
RAF Bridgnorth, September 1950 during basic training	48
214 Squadron, RAF Upwood, 1951	49
Peter with a Lincoln RE360, RAF Upwood, 1951	50
Peter fixing the propellers on a Lincoln aircraft, RAF Upwood, 1951	51
Working on a Lincoln engine at RAF Upwood, 1951	51
At RAF Upwood, 1951	52
At RAF Upwood, 1951, with John Wright and Dave Haylett	52
Ready for guard duty, RAF Shallufa, Egypt, 1951	53
Crossing the Sweet Water Canal by raft	53
With a water cart donkey, by the Sweet Water Canal	54
In the desert, RAF Shallufa, Egypt, 1951	54
Outside the tent, RAF Shallufa, Egypt, 1951	55
Looking spruce, RAF Shallufa, Egypt, 1951	55
214 Squadron, RAF Shallufa, Egypt, 1951	56
En route to the Canal Zone, Egypt, 1953	59
The NAAFI wagon is here! Egypt, 1953	60

Our billet in Kenya, 1954	64
At RAF Eastleigh, Kenya, during the Mau Mau emergency, 1954	64
On the way to do farm guard duty, Kenya, 1954	66
On the farm in Kenya, 1954	66
Farm guard duty, Kenya, 1954	67
Kenya, 1954 - the refuelling team	67
The Government mercenary, Kenya, 1954	69
RAF Eastleigh, Kenya, 1954: the writing on the wall	70
214 Squadron, Kenya, 1954	71
RAF service record, 1950-55	73
The 'Accountant' aircraft	78
Building the 'Accountant', at Aviation Traders, Southend airport, 1956	79
A short-nosed Bristol 170 Mark 31	79
A Bristol B170 Mark 32, Southend airport, 1958	80
A Bristol Freighter B170 Mark 22, Southend airport, 1959	80
Setting off from Southend to Australia, 1958	81
A DC4 G-APNH, Southend airport, 1963	81
A Bristol Freighter airborne	82
DC 3 Minor Maintenance Team, Aviation Traders, 1964	82
HeavyLift Belfast aircraft G-BEPS at Stansted	88
The Carvair G-APNH	91
DC8 aircraft, Stansted, 1979	103
CL44 G-AZIN, Stansted, 1979	104
DC 8-54 G-BTAC, Stansted, 1979	104
Belfast aircraft at St John's, Newfoundland, 1998	107
'Screech' bottle	109
At the Paris Air Show, 1983	116
On holiday in Brussels with Vera, 1984	116
At an air show in Czech Republic, 1998	117
Air show at Forli, Italy, 1996	124
Warsaw Air Show leaflet 62nd anniversary of Warsaw Uprising	126
The VIPs, Veterans and Sally B crew, Warsaw Air Show, 2006	127
Sally B dropping leaflets over Warsaw, August 2006	128
Peter dropping leaflets over Warsaw	129
Peter being presented with a PAF plaque	130

Sally B on the way to Buckingham Palace, VE Day 60th Anniversary	132
The B17 fly-past over The Mall, VE Day 60th Anniversary	133
Sally B flying over Buckingham Palace, VE Day 60th Anniversary	134
Last edition of 'The Flying Dutchman', May 1945	135
Peter holding a piece of lunar rock	136
VIP visitor Lord Trenchard, after a flight in a DH86B of 24 Squadron, 1930s	137
Sally B crew being presented to Prince Charles, Duxford, 2002	138
Peter talking to Walter Cronkite, Duxford, August 2005	139
Meeting Václav Havel, 1994	140
The Sally B on a memorial flight for USAAF, 2002	142
The 5000th B17 built since the attack on Pearl Harbor	143
Annual Flyover, Madingley War Cemetery, 2000	144
With an old Flying Machine and its pilot, Duxford, 2003	144
The Sally B	148
The Sally B B17G crew and team	149
Working on the 'Memphis Belle'	151
Peter's acting role during the making of 'Memphis Belle'	154
With the B17 crew, cup-winners at the West Malling Air Show	155
CL44 G-AXUL being loaded at Stansted, 1979	159
A helicopter being loaded on to the CL-44 Guppy, Stansted, 1989	160
Belfast being loaded with aircraft fuselage, The Netherlands, 1998	160
Satellite dish being loaded onto a G-BEPS	161
Bus being loaded onto Belfast aircraft, Stansted, 1999	161
Barges being loaded onto the CL-44 Guppy	162
Giraffes being loaded aboard a CL-44 Guppy, 1987	164
A giraffe aboard the CL-44 Guppy, 1987	164
Cartoon of Captain Keith Sissons and Engineer Peter Brown	165
On board Concorde Reg G-BOAC en route to Sweden, 1987	168
Sally B in formation with Concorde at 'Vasteras' Sweden	169
The B17G Reg G-BEDF flying over Concorde	170
In Singapore outside the POW chapel in Changi prison camp	171

In Barranquilla, Colombia, 1985	171
Winners of the West Malling Air Show Bob Richardson Cup, 1991	172
The winning team in colour, Stansted, 1991	173
A *dacha* in Ulyanovsk, Russia, where Peter stayed on one trip	175
Peter with Captain Webb, Russia, 1995	176
Outside Lenin's house in Ulyanovsk, Russia, 1995	177
Sightseeing, Russia, 1995	178
Outside the old Guard Room, RAF Upwood	180
Threshing tackle machine, Hillside Farm, Market Weston, 1939	188
Cutting the cornfield, Hillside Farm, Market Weston, 1940	189
Harvesting with a modern combine harvester	190
At Hillside Farm Museum with Vera and David Sarson	194
Ken in thought, Hillside Museum, 1999	194

Acknowledgements

Writing this book has given me immense pleasure, but without help from a number of people, it would not have been possible, so I thank everyone involved.

First of all Wendy Crozier of Bound Biographies for word processing and editing my manuscript, and for her patience in dealing with me over that time.

I would also like to thank the following for giving permission to reproduce certain photographs:

JHG Barrett for the map of the bombing in Hopton on page 30.

Terry Smart, Engineer, Aviation Traders Engineering for the aircraft photographs on pages 79, 80, 82 & 91.

Paul Brown for the 'Setting off from Southend...' photograph on page 81.

Simon Barker, Transmeridian Air Cargo, for the photographs of DC8s and CL44s on pages 103, 104, 159 & 160.

Steve Carter, Geoff Smith and Ken Reed, B17 Crew for the photographs on pages 124, 129, 140 & 141.

Elly Sallingboe, Operator of the only flying B17 in Europe for the use of the B17 Preservation photographs and for keeping that iconic aircraft flying for over 33 years and giving me the opportunity to be her Chief Engineer for the last 25 years. Pages 127, 130, 142, 144, 148, 149, 155 & 172.

Muzeum Powstania Warszawskiego for the photograph of the B17 over Warsaw on page 128.

Coert Munk and Captain Peter Kuypers Dutch B25 for the photograph on page 132.

Kevin Jackson for the photograph on page 133.

Ken Ellis and Jarrod Cotter, *FlyPast* magazine, for the photograph on page 134.

Dr Everett Gibson, NASA, for the photograph on page 136.

Aeroplane for the photograph of Lord Trenchard on page 137.

The Imperial War Museum, Duxford, for the photograph on page 138.

Cambridge News for the photographs on pages 139 & 144.

Boeing Aircraft Company for the photograph of the 5000th B17'on page 143.

Lord David Puttnam, Enigma Films Productions Ltd, for the photograph on page 154.

Peter Rooley, HeavyLift Cargo Airlines, and Peter Clark, Transmeridian Air Cargo, for the photographs on pages 161, 162, 164 & 173.

Peter Hale for his cartoon on page 165.

Kenneth Hudd for the photograph of Sally B in formation over Concorde on page 169.

Swedish Newspaper, VIT, for the photograph on page 170.

Nan Pollard, for the old farming photographs on pages 188 and 189.

David Sarson and Percy Prentice, 388th Bomb Group USAAF Museum, Hillside Farm for the photographs on pages 190 & 194.

A number of other people have given me permission to use their private photographs in this book, for which I thank them.

Aviation Traders Engineering Ltd., Channel Air Bridge, Air Charter (all Freddie Laker companies), Transmeridian Air Cargo Stansted, and HeavyLift Cargo Airlines UK are no longer in business.

I have made every effort to identify and trace all current copyright owners for their permission to reproduce the photographs in this book. If I have failed, I apologise in advance for any oversight in this matter.

My Childhood in Suffolk

My Childhood in Suffolk

To start at the very beginning, I was born on 9th November 1932 in Fen Street, Hopton, on the Norfolk/Suffolk border. I was the second son and fourth child of George and Annie Brown.

My dad was a farm labourer and worked at Hopton Farm. He had been in the Army (the Royal Garrison Artillery) for eight years, and had seen service on the Somme during the First World War, and also later in India during the 1920s. He married my mother after leaving the Army. My mother (née Smith) had lived in Thelnetham as a girl, but her father was at one time the farm manager at Hopton Farm.

Peter, age 5 (first school photo)

My first memories of living in Fen Street - and about the only two things I do remember - are watching the shoot wagon turning round by our house, and playing in the disused gravel pit. We moved to Market Weston in 1936 to Number 3 Council House, now called Number 3 Church Road. By this time our family had increased to six children, and was later to increase to ten, so even for that time we were a large family. I started school at Market Weston. My second sister was already there; the two elder ones had to stay on at Hopton as the school in Market Weston was only for children up to 11 years old. The rest of us went to Barningham at the age of 11.

In my early days village life was very simple. As children we played marbles, hopscotch and spinning tops on the road in front of our house; there was so little traffic around that it wasn't a problem. We made the marbles from clay and baked them in the oven. We did have the glass alleys sometimes, but when we played with one of those it was understood that you exchanged it for a clay one if you lost.

Market Weston School

It's not easy to compare those days with now, as so many things have changed. We were happy; it was an uncomplicated way of life with no pressure from the school and no homework. We had the odd exam but it was no big deal - we didn't even have school reports, and as far as going to university went, I think many of us didn't even know what one was. Even so, most of us made our way in life quite successfully. Maybe with a better education things might have been different - who can tell? I just can't make up my mind about it all; enough said!

Market Weston had a one-room school with one teacher, Mrs Carter; she was all you would expect of a village schoolteacher. Later on they built a wooden extension to the school for the infants. There were a few more children then, some being evacuated from London and my two cousins from Weymouth. They all lived in our house until they found a house in the village (a little crowded to say the least).

Most of the working class children had little jobs. My brother worked on the same farm as my dad and did odd jobs, collecting eggs and washing them (no battery hens at that time). I chopped wood and pumped water for an old lady in the village for sixpence (old pennies) a night.

During my childhood and the time before I left home there were many ways we could make a bob or two; I will try and give you some idea how that was done. I suppose the first was picking blackberries down on the Fen and selling them to a shop in Hopton, as I remember, the shop near the church.

The Fen was common land owned by the Parish and the local farmers could graze their cattle on it. The money they paid was given to the less well off at Christmas in the form of 5 cwt of coal and 7 shillings and 6 pence to buy food at the village shop. I can well remember Mum coming back from the shop on Christmas Eve with the shopping; you could get a fair amount for 7/6d at that time, or so I thought.

While I am on the subject of coal, there was another way to 'keep the home fires burning' (my mother's favourite song). When trees were being felled in the village by the wood merchants, the farmers would give the treetops to one of their workers to clear up and keep any wood that the wood merchants didn't want. That was a good deal all round, as the farmer got his money for the wood, we got our firewood, the land was all cleaned up by us, and everyone was happy.

We would pick up all the acorns that had fallen from the trees and take them to a small farmer for his pigs; he would mix them in with the other pig food. He didn't pay us much for them but, as we say, 'every little helps'.

One of the best money-makers was during the war years when the black market was in full swing. A man from London (or so we were told) would come to the village on a Sunday morning and buy up anything we had in the way of chickens, ducks, rabbits and eggs. It was not just hen's eggs he would take, but any type - even eggs from water hens as we called them ('moorhens' to other people). He would roll up in a big Ford V-8 car, so he must have been a real London spiv; but he paid well, so everyone was happy.

Most families kept hens and rabbit in the back yard, and some even had a pig, although we never did. We had several hutches with doe rabbits in them and a man in the village had a buck rabbit, so for a few pence all was taken care of. Rabbits being rabbits, there was always a good supply of them to sell. In our household we only had wild rabbit for dinner; we sold all the tame ones. Their rabbit food didn't cost us anything, as they were fed on what we picked from the sides of fields and lanes, such as hog-weed, clover, south thistle and the like; so you see, apart from paying the man with the buck, it was a low cost earner.

Wild rabbit was quite easily available as well. The boys of the village would go down to the Fen, or wherever else we could without being caught, with a ferret. The ferret was put down the burrow, we would place nets over the bolt holes, and then wait for the ferret to chase the rabbits out into the net. However, if the ferret decided to have dinner down the hole and then fall asleep, we had to dig it out, and that could take some time.

The best time for getting a few rabbits was during the harvest season, when the corn was being cut. We would go along the sides of the fields, and as the rabbits ran out of the uncut corn, you could catch them. On the last part of the harvested field, they would all run out; some were caught but some weren't, and the lucky ones lived to breed another day.

Now the system was that at the end, when the field was all cut, all the rabbits that had been caught were laid out and the headman or foreman would share them out. He would start with the best for himself and work his way down through the ranks. We boys were last in the line and sometimes the rabbits were all gone by the time the headman got to us, or all that was left would be the old or small ones. But the village boys were up to that game, and if we caught a good, young, big one, we would hide it under a sheaf, return later, pick it up, hide it under our coat and take it off home to Mum for dinner next day. The rabbits were valuable not only for eating - you could even sell the rabbit skins to the gypsies who came to the door selling pegs and wooden flowers.

One of the families in the village had what I think was the ultimate way of making some extra cash. They had an allotment and on part of it they grew sugar beet. During the sugar beet season, when the lorries were taking the beet to the sugar factory at Bury, some of the beet would fall from the lorries (not so much Health and Safety in those days). They would pick these up and take them home to add to their own crop. At the end of the season they would send them to the factory; the crop may have only been a part-load, but it was well worth doing.

One of the most organised money-earners was potato picking (I say this because it was organised through the school system). The reason was that during the war there was a shortage of labour and when jobs came up that required extra workers over the normal man/womanpower, that's where we schoolchildren came in. The deal was that we could have twenty half-days a term to help out on the farms, although I can only remember doing it in the potato picking season.

Once we got to the field we were given a basket, and the rows were divided between the number of pickers, so you all had the same length to pick. As the machine came along and lifted the potatoes out of the ground, you picked them up as fast as you could into the basket, emptied it into a sack, and then waited for the potato-lifter to come round again. If you were too slow and the lifter came round before you had picked the last round up, you were in real trouble, as once you got behind, it was a horrendous job to catch up with the machine again.

You had a card from school which the farmer would fill in at the end of the day to keep a record of the days you worked. It was a very good system: the farmer got his potatoes harvested, we got some time off school, and the big plus was that we got paid - not bad at all.

So those were some of the many ways that you could make some money in my school days. I don't think pocket money was around at that time (well, certainly not in our household). And that's the way it was.

There is one more little escapade I must recount; this one didn't earn us any money but it did earn us a little grief. It all began when my brother Denny and I were up at the American Air Base, Knettishall, towards the end of the war. Some 'GI's (as they were known) came up to us and said, "Hey you boys, would you like a goat?" As we were not used to saying no if we were offered anything, and not even thinking if we really wanted a goat or what our mother would say if we took it home, we said, "Yes please." And with that we were now the proud owners of one rather sad-looking goat.

Now to get it home we tied a piece of string around its neck and off we went. It wasn't a great distance to our house, but with a goat that didn't want to walk too fast or indeed not at all some of the time, it took us a fair while. We took turns in the goat-herding task and made steady progress home. But as soon as our mum saw what we had on the end of a piece of string, she went mad. The last thing she wanted was a goat. I won't tell you all she said, but in plain English it was to the effect, "Take that scruffy-looking animal back where you got it from!" The poor goat: the Americans didn't want it, and now it wasn't welcome at No 3.

What now? We couldn't hide it at the bottom of the garden with the rabbits and the chickens, and we couldn't take it back. No one else wanted it as a pet, so we had a problem on our hands. It was then that we came up with a cunning plan.

At the other end of the village was a smallholding called 'Weather Cock Farm' and the Mr Rush who lived there kept a few goats. So once again with goat in tow, we set off on the second goat-herding of the day. We didn't even know whether he would want a strange looking goat in his small herd, but we pressed on and after a while came within range of the farm.

It was then that we came up with the second cunning plan of the day. The reason for this was that we couldn't take the chance of the goat-man saying, "No - take that thing away!" Once we got near the farm we crept along on our hands and knees with the goat still on the piece of string. When we reached the farm gate we made sure no one was around, opened the gate very slowly so we didn't make any noise, untied our by-now friend, pushed her inside ("Goodbye goat"), shut the gate, and you never saw two boys run so fast.

I expect Mr Rush always wondered where that goat came from, but our goat was with other goats in a good home. Now in our case the moral of the story is, "Do look a gift goat in the mouth, and say 'NO'."

The biggest change that came to village life was the outbreak of the Second World War. Overnight, or so it seemed, airfields were being built, Army camps were springing up everywhere, and convoys of Army equipment were rolling through the countryside.

During the two years from 1937 to 1939, airfields were being built all over East Anglia; my dad worked at RAF Honington after being laid off the farm. Just prior to this it was a very bad time for employment, so the building of airfields was a great boost to employment in the area. It was the Irish from Southern Ireland that made up the main labour force. They built the accommodation for the airfields first, and then lived in this until the airfield was complete.

My eldest sister worked at the canteen during the building of the Knettishall camp, and once the Americans moved in, she worked for the American Red Cross. I remember that it was during the building of the airfield that we saw our very first film. They used one of the huts for the film show, setting the projector up on a table and changing reels as required.

Around this period, the war was obviously the main concern that preoccupied the population, and we were bombarded by the Government with propaganda slogans: "Dig for victory", "Walls have ears" and "Is your journey really necessary?" were but a few. One of our duties was to collect salvage for the war effort: pots and pans, old tyres - just about everything was collected. We (that is the young boys of the village) would meet at Dr Newham's house on a Saturday morning with old prams, handcarts and the like, and then go around the village collecting.

Dr Newham was also head of the Home Guard, or as it was in the beginning, the LDV (Local Defence Volunteers). My dad was one of the first to volunteer for the LDV; I think he would have liked to join the Army again but by that time he was probably too old. Then the LDV became the Home Guard and men of a certain age were forced into it. My eldest brother Jack was in the Home Guard before being called up for the Army, going over to France on D-Day and then on to Germany until the war ended. After that he was released as 'A-class' to work back on the land, as at that time, with Europe to feed, there was a great shortage of food, even worse than when the war was on.

Going back to those wartime days, the biggest change in village life came with the arrival of the Americans. They had money to spend and all the food they could eat, things we hadn't seen for years. As children we were allowed into the camps to visit the cinema or the gym and play ping-pong in the Red Cross canteen, drinking *Coke* and eating doughnuts; this was a new world. At times we would even eat in the dining hall. I don't think this was really allowed but nevertheless we did it.

The Americans would buy up all the bicycles that were available (to get a new one you had to have a Government Permit). Once they had bikes they would visit all the public houses and by halfway through the delivery, the pub would run out of beer. A sign outside would say 'Cider only' but often it was a case of the door being shut and the back rooms being full of Americans drinking beer at the right price. But I must say that despite any resentment this might have caused, that period has always been remembered as the time when life came to the village.

The Market Weston salvage collectors; Peter is seventh from left and brother Denny is first on the left

Life in our house changed during the early part of the war. With Dad and Jack in the Home Guard, we had their rifles and their uniforms in the house. My dad, being an old soldier, did not think much of the men in command of the Market Weston troops; they were the gentry of the village. He was asked once while on parade, "Brown, what is the first thing you do before cleaning your rifle?" The expected stock answer was to make sure that it was not loaded, but no, back came his reply, "Check the serial number, making sure you are not cleaning some other bugger's rifle!"

We all had gas masks, ranging from a large one that you could put the baby of the family (we always had one) in completely; it had a sort of bellows on the side to pump air in. The younger children had the 'Mickey Mouse' type; they were red and blue with a sort of nose on the front. The rest of us, apart from the Home Guard two who had regular Army-type ones, had what was the standard civil type; they came in a cardboard box with a string attached, which was for carrying around your neck. We had to take these with us at all times, although apart from school I don't think we took them anywhere else. We did have a gas mask drill at school which involved pulling them on and holding a card at the filter, then breathing in and making sure the card stopped there until you breathed out.

The schoolhouse windows were all painted with a solution to prevent the glass shattering during an air raid. The only air raid shelter we had at Market Weston was a tabletop over the desks, which we only got under once when the Germans bombed Hopton one afternoon. At Barningham School we built an air raid shelter by digging a big hole with a timber and tin roof covered over with earth. The only problem was that the rain filled the hole and there was no pump to pump it out. It was all right during an air raid if you could swim!

Market Weston did get bombed and machine-gunned. The bombing was one afternoon; I was at school in Barningham and we heard it. There were a few stories about the damage, but when we got home the string of bombs went across the road from just at the back of our house to a field at the back of the Post Office yard. One bomb was in the centre of the road and one knocked down a row of buildings just opposite the Post Office. No one was hurt.

MY CHILDHOOD IN SUFFOLK

Barningham School, 1943; Peter is 6th from left and Denny is 2nd from left, next to the teacher

FROM PLOUGH TO PLANE

Map showing the bombing of Hopton, November 1940

MY CHILDHOOD IN SUFFOLK

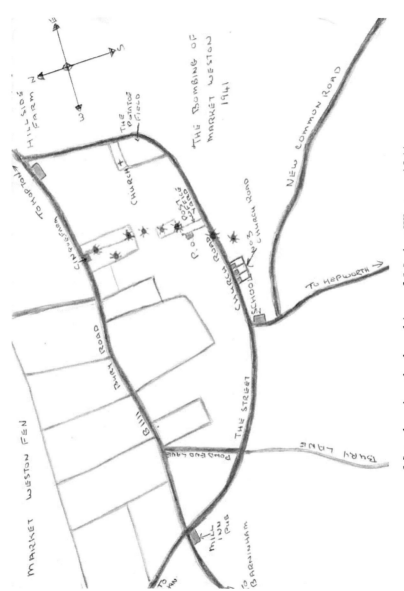

Map showing the bombing of Market Weston, 1941

The machine-gunning happened late one afternoon; I was in the back yard with my mum; she was paying the coal man and this plane came over. We watched it until we realised it was a Gerry and that he was machine-gunning the cows in the field.

Most villages did get bombed. At Hopton one Saturday night my dad was in the Greyhound pub when the house next door got bombed flat. The people there were evacuated from London to escape the bombing and ironically got killed in the countryside instead. Barningham had a stick of bombs in the fields, but they came from a B17 flying out of Knettishall airfield.

The war also introduced the population to rationing. Everyone was issued a ration book for food and clothing. I don't think this had any great effect on the Brown household as money was the thing on ration as far as my mum was concerned. The clothing coupons my mum sometimes sold for some extra cash. I can remember taking them to one of the richer families when my mum was a bit short of cash.

As the war went on it seemed to us that it had gone on for ever. My dad used to tell us about the trenches in France: the mud, the shelling, the cold, wet conditions, and fleas and rats everywhere. He also told us tales of his time in India. I wish I had taken more notice of him but at that age it seemed like a hundred years ago. Looking back now of course, I realise that it was a very short time between the two world wars.

When we moved to Market Weston my dad's mates were still in Fen Street and sometimes he would go to visit them on a Sunday. He would take me with him on the crossbar of his bike and while he was talking to his mates (these were men he had been in the Army with) I would be given a *London Illustrated* magazine to look at; these were also full of First World War pictures.

My dad went back to work at Hopton Farm after the short time he spent at Honington airfield. One of his jobs was to collect the waste from the mess halls of the American camp to feed to the pigs. This was done with a horse and tank on wheels (it was really a water cart). There were two mess officers and other ranks. I would sometimes go with my dad and I remember one mess sergeant getting mad because the horse, being a creature of habit, would always leave his calling card right outside the officers' mess.

MY CHILDHOOD IN SUFFOLK

In 1943 my brother Jack was called up for the Army; he went to Bury St Edmunds' Gibraltar barracks for his basic training and then to Teddington, London. That is where our Uncle Sid lived, my mother's brother; he had served in the First World War and also in France. Uncle Sid had been a gardener in Norfolk but had somehow moved to London to become a gardener at one of the big houses.

The war rolled on, with some of our local lads being held in prison camps in the Far East after the Japanese took Singapore. My Brother Jack went over to Europe with the D-Day invasion; he was in the Pioneer Corps and spent most of his time laying airfields. It was during his time in Europe that we had one of the saddest days in the Brown family. One Sunday morning (the 6th June 1944 - D-Day) our young brother Brian, aged five, was knocked down by a lorry outside our house and killed. Tragically he didn't hear the lorry because there were so many planes overhead landing and taking off from the local airfields on that momentous day. My mum didn't tell Jack at the time, as on D-Day he was going over to the Normandy beaches for the landings with the Allied invasion forces. Understandably, she thought he had enough to get on with. However, he soon wrote to say he would be home on leave and so she had to write to him. He said he knew something was up, as my mum hadn't mentioned Brian in her letters to him for a while.

Jack came home on leave and on his last night before going back he gave my mum £20, which was really a lot of money for us. He had been paid for his time in France and Germany and since he had not spent much over there, he had quite a lot left. With the money, my mum bought a radio set from Curry's in Bury St Edmunds; it was a 'utility' model with Home and Forces programmes.

The radio had to be battery-driven, as we didn't have electricity until well after the war, when I had left home. To run the radio you needed three batteries - an HT 120 Volt, a grid 9 Volt and a rechargeable accumulator 2 Volts. The accumulator was charged at the local garage for about tuppence. The other two batteries were expensive, so it was a bit of a disaster when they ran out.

Not only did we have no electricity but there was no water either, apart from the pump in the front garden to serve the six council houses. When we moved in, we found that the pump was a strange affair; a chain went down to the bottom of the well, round a pulley, with the same arrangement at the top. You turned the chain by means of a handle turning the top pulley. The chain was made up of small buckets that went down the well upside down; on reaching the bottom they would fill up with water and, being the right way up, would carry the water to the top. As the chain went over the top pulley, water would fall out into a chute and into your pail hanging on the outlet spout. So it was a rather complicated way to get the water up from the well, and the chain was always breaking, which meant no water; then we would have to beg the house over the road for a pail of water. This was also the case in the very hot summer when the well ran dry.

When the chain broke, the council man would come with his grappling tools and fetch the chain up for repair. Later on we went to the traditional well with a bucket and rope, which was a bit more dangerous with lots of children about. I don't think we lost any children (although maybe a cat or two) but it did away with the breaking chain problem.

The clothes washing water came from the pit opposite our house. Our washing was done in the copper in the kitchen. Water from the pit was also what the steam engine pulling the threshing machine would fill up on.

There is another story I must recount here regarding the pit. During the war, aircraft carried as part of their emergency equipment in the dinghy, a tin of Sea Marker powder. This was so that if they ditched in the sea and were able to get into their dinghy, they could sprinkle the powder overboard and it would turn the sea a brilliant green, which meant that they could be seen from the air by the rescue aircraft. Now we children found a can of this up at Knettishall base and took it home. On the way back we tried it in a few puddles of water and it worked fine. So when we got home I thought it would be a good idea to try it in the pit - not such a good idea. We discovered that not only did it turn that pond a brilliant green but it found its way via the drains into numerous other ponds too. So as you can guess, that was not the best idea I ever had!

Apart from the short time my dad worked on Honington airfield, he was always a farm worker. Farm work was very hard in my early days and men worked six full days a week during some seasons of the farming calendar. Some of the labour was done on piecework, which meant that the harder they worked, the more money they could earn. One of these jobs was, as it was known, 'chopping out'. That was hoeing the sugar beet when they were small plants. The task was to leave a plant every foot or so along the row so that the ones left could grow into big sugar beets. To speed up the hoeing and earn a bit more money was where us young ones came in. This job was known as 'singling'; what you had to do was go behind your father and when he left more than one seedling you had to pull out the small ones and leave only one plant. This was so that Dad didn't have to waste time fiddling with his hoe to separate the small plants. As a reward for the work he would give you a penny or so, but those fields in Norfolk were big and the rows were very long; however that was the late 1930s and times were hard.

One of the big advantages in doing a good job first time round and not leaving too many plants was that when the second hoeing took place (which was called 'scoring up'), you could do it quickly. This meant you could earn more money, as you did the same rows that you had weeded the first time on your second time round.

One of the jobs I had when I first started work was horse-hoe leading; let me just explain what that was. The sugar beet and the like are in sown in lines and have to be weeded, but they can only be weeded between the rows. That was done by a horse-drawn hoe doing about six rows at a time, with a man guiding the hoe and a boy leading the horse between the rows. That worked fine most of the time, but now and then the B17s, or any plane come to that, would cast a shadow. As the shadow went over, the horse would take fright, and then it was hard work controlling it - big horse, small boy.

While doing this work on one of the fields near a small farm, I saw they had a notice saying 'goslings for sale'. So I thought I would buy some, keep them until Christmas, get them nice and fat, and make a bob or maybe two. That is what I did and yes, at Christmas I sold them and made a small profit; the only problem is that geese are friendly birds and I was sorry to see them go! But then that is the way with village life; it always was and still is today.

I had other livestock projects too. I can remember catching the bus to Bury and buying day-old chicks at the auction on the cattle market and taking them home on the bus. They were mainly cockerel chicks, which meant they didn't cost much but having cockerels was fine with me.

I had made a home for them out of a wooden box heated by a paraffin heater, which I bought for the job of keeping the chicks warm during the early stage of their lives. The only place suitable to keep them was in the kitchen. I don't know what my mum made of it, but she let me keep them there until they were old enough to go outside. I would check them during the night; if they all sat round the lamp without making a noise, they were happy. Then when they could fend for themselves, I sold most of them, keeping about ten or so to fatten up for Christmas.

Peter's two geese outside his home in Market Weston, 1947

Before I left school at age 14, I worked on a small farm in Market Weston at weekends and in the holidays, where I did the milking (no machines) and general farm work. One of my jobs was to look after the POWs (prisoners of war) - we had a prison camp nearby and the prisoners worked on the farm. At one time we had Ukrainians, and at other times Germans or Italians. Some of the POWs stayed in the village on the farms where they worked, but this only applied to the Italians; the rest of them had to go back to camp each night.

I continued working on that farm after I left school. The only other worker was a lad from Hopton (Peter Savage). We got paid on a Saturday morning but the boss was always short of cash. So before we could get our wages, we had to kill and pluck a chicken or duck, which he would take down to the butcher in Hopton to get the cash and then pay us. Otherwise he would give us a cheque that we had to try and cash - not easy in Market Weston in the 1940s, and not that easy today either.

I remember once on a Tuesday, the farmer and his wife had gone to Ipswich market. That was also the day that the bus ran from Hopton to Thetford for the midweek pictures, so we decided to make the trip, but we had to milk the cows first. We decided that after dinner we would get the cows milked and then go home early, get ready and catch the bus. However, the cows, also being creatures of habit, were not ready for milking, so we only got about half the normal amount. Now, every day the farmer would check the amount his cows were producing. When he came home and checked the level of the milk churn, he was furious and rushed round to our house before we caught the bus to Thetford to find out exactly what was going on. We were certainly not in his good books for a few weeks after that.

Another job I had when they both went to market was to make tea for the POWs working on the farm. I had to boil the water in the farm kitchen, which had a paraffin stove. On one occasion I filled the kettle (a large one), went outside for a while and when I came back in to make the tea, you never saw such a mess! I had put the wick on the stove too high and it had smoked the whole kitchen out. There were black cobwebs everywhere; it looked like a witch's kitchen. I had to spend the rest of the day clearing up before they came home.

FROM PLOUGH TO PLANE

I left Hillside Farm in Market Weston after a dreadful back-breaking job I never wanted to do again. The boss decided that we two boys working there should dig up a field of potatoes by fork. It was a four-acre field - not that big for a field, but very big for just two boys to dig up. Moreover the ground was very hard and the potatoes had not been planted in the conventional way, but using a plough, so the land was flat and not in the usual ridges you would expect.

After that I had a job in the village working for someone who had a large garden and a goat, and bred dogs. I milked the goat and did the gardening. Next I got a job in the next village, Coney Weston, on a large farm with no cows but stock cattle and sheep. I had many jobs on this farm, ranging from looking after a yard of bullocks to rook scaring, stone picking and working as a shepherd boy.

By then the war was over. My brother was back from the Army and he worked at the same farm as myself. Being a man he worked with the main gang and quite a lot of the major jobs were on piecework. Being only a boy still at this time, 16 years old, I worked with the horse man and the shepherd and drove the tractor (at that time no one was too worried about driving licences). One of the jobs with the tractor was getting the airfield back to farming land, as a lot of the land the airfield took was from the farm where I worked.

The job with the shepherd, a big man called Bert, was quite interesting; I would help with the lambing and shearing. There was no electricity on the farm at that time, so when we did the shearing, my job was to turn the handle on the gearbox, which in turn drove a flexible driver to the clippers. The shepherd would shear the sheep after I had been into the pen and caught one, so he had a short rest between one sheep and another. In contrast, I was kept constantly busy with turning the handle and catching the sheep - and to top it all he was paid one shilling per sheep as a sheep bonus. Oh yes, I forgot to say that after he had finished shearing one I had to roll the fleece up in a bundle and stack it in a pile. The only plus was that after a few hundred fleeces to bundle up I had really soft hands with the oil from the sheep's coats.

One of the jobs with the shepherd was to tail the lambs. This was done by a hot type of cutting tool. We used to build a fire in the field and keep the tool at the right temperature. The other job was to take care of the boy lambs' bits; I don't think I will go into that right now!

I worked on that farm until I decided that maybe farm work was not for me. I remember one Saturday night I was cycling to one of the villages to see some mates and halfway there I had to take shelter from the rain under a tree. It was pitch black, cold, and I was very wet and that was, in my case, the final straw. I had mates in the Merchant Navy, so I thought that's what I'd do, join the Navy. That very night I turned round, went home and sent in the 'Join the Navy' advertisement from the News of the World newspaper. I didn't tell my mum and just waited for the reply.

When the reply came, it turned out that the shortest time you could join the Navy for was eight years. When you are 17, eight years seems likes half your life. The RAF on the other hand let you join for a five-year stint, with four years in the reserves. So I sent off another application, this time using the 'Join the Royal Air Force' advertisement.

So began my next step. I got a letter back with a travel warrant to go to Norwich for a medical and IQ test. I got a ride into Thetford, our nearest rail station, with a man from Market Weston who worked in Thetford and from there caught the train to Norwich - now, that was my first-ever train ride at 17½ years old.

I passed the medical and the IQ test and waited for the next move. Meanwhile I gave my notice in at the farm, and very soon I received my instructions from the Air Force to report to Cardington; and so began my Air Force days. This was now September 1950.

The weekend before I went to Cardington all the lads had a few beers at a pub in Blo Norton, *The Case is Altered*; I think it is long gone. I remember it was a good night. I sold my watch and also a kipper tie I had, with a picture of a girl under a palm tree on it. So with about four or five pounds in my pocket, off I went into the unknown. My mum cried the morning I left home. Was it with joy or sadness? I don't know for sure, but I think it was the latter!

Peter outside the house where he was born, in Fen Street, Hopton

All the family with Mum, Market Weston Village Hall, 1980

Keeping my hand in on the tractor!

My Air Force Days Begin

My Air Force Days Begin

So after saying goodbye to my mum I got a lift into Thetford and a train to Norwich to meet up with the rest of the boys from around Norfolk for the journey down to Cardington. There were about ten or so of us on leaving Norwich but by the time we had reached Cardington, some had decided that maybe a life in the Air Force was not for them and had got off the train at a stop between Norwich and Cardington.

Cardington in those days was the camp all Air Force recruits went to, or at least all those who had volunteered, as at that time we in the UK still had conscription and all young men that were eligible were called up. As I worked on a farm, which was then classified as an essential job, I would not have been conscripted, although in later years, after my age group, some of the village lads were called up. Much of the same happened during the war, when some of the farmers' sons got considered to be on essential war work by calling themselves farm managers.

We spent a week or so at Cardington being kitted out with uniforms and a full set of what was called webbing; that set of kit consisted of a big and small pack to carry your belongings. The big pack was carried on your back and the small pack over your shoulder. There were also two ammunition pouches strapped to your front and a water bottle which was carried by the webbing belt that you wore around your waist. This set of kit was completely useless and served no practical purpose whatsoever. You had to keep it clean and lay it out for a kit inspection whenever they came along. The only time you wore this kit was if you were on jankers, which was the punishment for getting into trouble with the Air Force rules, such as being back late from weekend leave, having not had a haircut for the last few days, or anything else on a long list of misdemeanours. What you had to do was, after work, get dressed

FROM PLOUGH TO PLANE

RAF Cardington, 1950 (first photograph taken in the RAF); Peter is on the right-hand end of the front row

MY AIR FORCE DAYS BEGIN

up in 'best blue' (your number one uniform) with all this webbing, and report to the guardroom. Here you were inspected by the orderly officer to make sure everything was clean. Then you did some manual task, for example, polishing the guardroom floor or whatever they could find for you to do.

Cardington was fine; no shouting or rushing about - that came later. You were sworn in at Cardington to serve your King and Country. So once that was done the next stage was Bridgnorth in Shropshire.

A special train was laid on to take the next intake to Bridgnorth, as that was where we would do our square bashing. All went well until we arrived at Bridgnorth station and then all hell was let loose; we didn't know what had hit us. DIs (drill instructors) were shouting at us and throwing our kitbags all over the place. It was then I thought that those who jumped the train between Norwich and Cardington maybe knew more than the rest of us. At square bashing the drill instructors were God Almighty and for the first few weeks we didn't have time to breathe. From early morning to lights out we were running, marching, polishing floors, cleaning kit and trying to get our boots shine like glass using all the handed-down methods. We used hot spoons to get the toe-caps smooth and spit and polish to make them shine.

During our eight weeks' training, also with the threat of being put back a week to the next intake for not coming up to standard, our flight was picked to go to London to line the route for the Dutch royal family on a state visit to England. We stayed at RAF Uxbridge during this visit, which was rather a change from Bridgnorth. It made our eight weeks a bit shorter and nearer our passing out parade, which once it came was rather rewarding; you felt on top of the world, an Airman at last, although be it only AC2 (Aircraft man second-class), the lowest of the lot. After the passing out parade was over we then had a posting to report to.

Following our leave almost all the men went to training camp for their trade training but some reason I was posted to an operational station. This was a new scheme called 'on-the-job training'. I don't think it lasted long as I never met anyone else who had done the same type of trade training. I was going to be an engine mechanic, liquid cooled, as the station I was going to do my training at, RAF Upwood, had four squadrons of Avro Lincolns, Nos 7, 49, 148 and 214.

FROM PLOUGH TO PLANE

RAF Bridgnorth, September 1950 during basic training (square bashing); Peter is in the middle row, second from right

MY AIR FORCE DAYS BEGIN

214 Squadron, RAF Upwood, 1951; Peter is in the back row, sixth from right

I went home on leave after square bashing and then reported to Upwood. There were only three of us for the 'on-the-job training', one from Dudley in the Midlands, the other from Taunton, Somerset and me. In fact I think they didn't quite know what to do with us - three airmen with only eight weeks' service on an operational base.

After we had 'arrived', as they called it, we were passed to ASF Aircraft servicing flight. They did all the major inspection on the Lincolns. After a while I went to the Seven Squadron until they went overseas, then back to the ASF and a small amount of classroom training.

Peter with a Lincoln RE360, RAF Upwood, 1951

MY AIR FORCE DAYS BEGIN

Peter fixing the propellers on a Lincoln aircraft,
RAF Upwood, 1951

Working on a Lincoln engine at RAF Upwood, 1951;
Peter is on the left

At RAF Upwood, 1951; Peter is on the left

At RAF Upwood, 1951, with John Wright and Dave Haylett

MY AIR FORCE DAYS BEGIN

Ready for guard duty, RAF Shallufa, Egypt, 1951;
Peter is on the right

Crossing the Sweet Water Canal by raft

FROM PLOUGH TO PLANE

With a water cart donkey, by the Sweet Water Canal

In the desert, RAF Shallufa, Egypt, 1951

MY AIR FORCE DAYS BEGIN

Outside the tent, RAF Shallufa, Egypt, 1951

Looking spruce, RAF Shallufa, Egypt, 1951

FROM PLOUGH TO PLANE

214 Squadron, RAF Shallufa, Egypt, 1951; Peter is 3rd from right in the middle row

MY AIR FORCE DAYS BEGIN

They then decided that in the New Year, 1951, more recruits would be joining the same training and we were posted once again to ASF to wait for the new recruits to join us. I spent some time in the tool store handing out special tools to the men working in the hangar.

I was still reading the training books and it was during this period that 214 FMs, 'Federated Malay States' were going on a detachment to Egypt, to a place called Shallufa in the Canal Zone. They were short of two mechanics so the flight sergeant in charge of ASF gave me a quick technical exam and I became a fully checked out engine mechanic with the great rank of AC1 (Aircraft man first-class).

I reported to 214 Squadron with the lad from Dudley and then with the rest of the squadron prepared for the detachment to Egypt. The exercise at Shallufa was named 'Sunray'. Most Bomber Command squadrons took part in the Sunray training at some time.

We got kitted out in our khaki drill; that was the dress we wore in the desert. There were shorts for the daytime and long trousers for night-time, all khaki in colour.

We left Upwood in an Avro Tudor; at this time the Tudor had been withdrawn from civil flying service because there had been some gone missing when flying around the Bermuda Triangle. This is an area of sea close to Bermuda where there had been some strange happenings with aircraft and ships. Soon after these events, the trooping flights were stopped and the aircraft became freighters. I met up with these aircraft later on in my career as an aircraft engineer in civilian life.

Our first stop was in the South of France for fuel and a meal. This was the first time I had been overseas in my life and I just couldn't believe it. From a farm in Market Weston to the South of France on my way to Egypt - what a change in a few months. I was 18 years old.

We took off from France; next stop Malta for a night's stop. I had palled up with two other engine mechanics, John Wright and Dave Haylett. Both were ex-boy airmen and had done 18 months in the boy service, so we were quite different in our service life but were good friends. This was 1951, and in fact, through the television Ceefax service 'Pals Search' we met up with each other again in 1998, some 47 years later.

Next stop and our final destination was the Canal Zone, Egypt, and what a change - desert sands and sun. The accommodation was in tents, with three to a tent, so the three of us were in the same tent, which was all right, as we were all new to this. The floor of the tent was just sand with a duckboard by your bed, and on the centre pole was a clay water container.

We had a local man, or as they used to say, 'your boy'. He kept the tent tidy, raked the sand and filled the water container up. It was quite an adventure for us; the only bugbear was guard duty at night. Being a 'Brown', and given that they always started in alphabetical order, I was soon on guard duty of two hours on guard with four off. The plan was to get on the first duty, which was 18.00 to 20.00 hours. The next slot was midnight until 2 o'clock in the morning, and then you were finished and resting until 06.00 hours, so you had four hours off. If you were on this last duty you didn't finish until 06.00 hours but the old hands seemed to somehow get on the first duty. It was also very cold at night in that part of the world so you had to wear your greatcoat on guard. You wouldn't believe, by contrast, how hot it would get in just a few hours.

Shallufa was alongside the Suez Canal and you could walk to it from Camp. You had to cross what was called the Sweet Water Canal. It was built as a supply waterway during the building of the main canal; that is where all the locals who lived alongside did their washing, washed themselves and took care of all the other requirements of life. You crossed the canal on a raft made of oil drums and were pulled across by a local for a few ackers. It was said that if you ever fell in you would need all the jabs in the sickbay to fend off all the illnesses you would catch.

During this stay in Egypt I went on a training flight to Aden. Now at that time Aden was a duty-free port and you could buy almost anything there. Back at home we had not got over the wartime shortages, but there they had watches, cameras and toys in abundance. It was great fun haggling with the traders; you always thought you had a really good bargain, but they were never the loser and most of the time you found someone got the same thing cheaper - or so they said.

It was interesting along the canal watching the ships sailing through and going into Fayed - that was the nearby town and had the big

MY AIR FORCE DAYS BEGIN

NAAFI club alongside the Bitter Lakes where you could swim. They were very salty so that you could almost float.

At this time King Farouk was still the Egyptian leader and things were quite relaxed. When I went back a couple of years later, the King had been removed, the Army was in control and troops were everywhere. One day we went over the canal as we had done on my last detachment, and being the old hand at that time, I didn't realise how much things had changed. As a result I nearly got us shot by not taking too much notice of the Egyptian army stationed alongside the canal.

En route to the Canal Zone, Egypt, 1953 (an Avro Tudor in the background); Peter is second from left

We returned to Upwood as men of the world, or so we all thought, and carried on with Squadron life on 214. As I have said, we did another detachment to Shallufa later on in my life path, but by this time things had moved on. The second time we were in huts, but I thought tents were better. The reason for the Sunray exercise was to take part in bombing practice, plus a competition between the Bomber Command squadrons.

Our next detachment was to Kenya to take part in the operation against the Mau Mau terrorists.

FROM PLOUGH TO PLANE

The NAAFI wagon is here! Egypt, 1953

214 Squadron in Kenya: 1954

214 Squadron in Kenya: 1954

During the emergency in Kenya against the Mau Mau, I was stationed at RAF Upwood on 214 Squadron (Federal Malayan States).

We had eight Avro Lincolns on our squadron. Apart from the odd exercise in the Suez Canal Zone, Egypt (described in the previous chapter), when we went to RAF Shallufa for bombing practice on what was called 'Sunray', we mainly stayed put at Upwood.

There were already some Lincolns in Kenya when we were told that our squadron would be going, and the long dash began to get our aircraft ready for the flight to Kenya. We left Upwood a week or so before the Lincolns in York belonging to Skyways. After a good night in one of the Ramsey pubs and a final cup of coffee in the Copper Kettle we left for RAF Eastleigh in Kenya.

When we arrived at Eastleigh which was on the outskirts of the capital Nairobi, the camp was well overcrowded. On the first night after we had drawn our bedding we had to find a bed space in one of the permanent billets. Most of us found a space in the centre of the billet floor between the rows of beds; not the best start after travelling all that way in the old York.

Next morning it was decided the best place for us was now the overcrowded gym hangar. When we arrived in the hangar we found the RAF Regiment, the Royal Signals, and the Harvard ground crews had already laid claim to most of the floor space. They had formed themselves into what I can only describe as individual settlements. We tried to do the same but it wasn't long before the old 'chiefy discip' was there and our beds, duckboards and mosi nets were lined up in a style the guards would have been proud of.

FROM PLOUGH TO PLANE

Our billet in Kenya, 1954 (the sports hangar had seen better days)

At RAF Eastleigh, Kenya, during the Mau Mau emergency, 1954

214 SQUADRON IN KENYA: 1954

After we had got the accommodation sorted out the aircraft arrived, only four at first. So there we were in a few days from the quiet dispersal of 214 Squadron at Upwood village with the Cross Keys pub across the road, to East Africa with guard duties and the hangar walls sometimes lined with Mau Mau suspects. How things had changed!

One of the strangest things about Eastleigh was that it was still the international airport with passenger airlines landing during the day: BOAC with the Argonauts, Britavia with Hermes, Air India with the Constellations, plus Vikings and the DC3s. So on one side you had the civil side with the international arrivals and departures and over on the military side, the active bombers and the famous Harvards all ready for the next raid - rather a strange set-up.

The Harvards' 13.40 flight had come up from Rhodesia and had their aircraft modified to carry eight 20lb bombs under the wings and had guns fitted (Browning machine guns with several hundred rounds of ammunition). We on 214 did not have much to do with the Harvard flight; as far as that went we played them at football and cricket, but I did fly in one of their aircraft.

One day our CO came along and asked for two volunteers to go upcountry as farm guards. I was not in the habit of volunteering but this sounded all right. So next day after I had got my small pack ready, gone to the armoury and drawn out a 303 Lee Enfield rifle and bandoleer of ammunition, we were ready for our flight to Nanyuki, which was north of Nairobi. Not to waste the taxpayers' money, the flight to the farm was made after a strike against the Mau Mau in the Aberdare Forest Mountains. The system out there was for the ground troops to radio the Kenyan police where the terrorists were. The Kenyan police, in their tripacers, would drop marker flares and then the Lincolns or Harvards would drop their bombs into the forest. So that was how I arrived to do my guard duties. My partner for the operation was a radar mechanic.

The reason the farmer needed someone to guard his family was that he and his elder daughter were to be in Nairobi for two weeks, which left his wife and younger daughter on their own at home. I enjoyed the two weeks, living in the farmhouse and roaming the plains. We went hunting with our 303s, but I wasn't much good at being the big game hunter.

On the way to do farm guard duty, Kenya, 1954; Peter is in the front cockpit

On the farm in Kenya, 1954

214 SQUADRON IN KENYA: 1954

Farm guard duty, Kenya, 1954

Kenya, 1954 - the refuelling team

We had to do guard duty at night, with the RAF Regiment doing it during the day; but the fitters, riggers and, as we called them, the ancillaries doing all other trades. Doing the night shift never made much sense to me. However, that's how it was, and we did have the soldiers of the King's African Rifles to help us. The aircraft did a lot of flying and the hard work was for the armourers, as it only took a short time for the aircraft to drop their bombs and be back on the ground for another load.

We did have one fatal mishap while on a low-level bombing sortie with aircraft SX976. It is believed that the bombs were fused with instantaneous inpack rather than the delay inpack and a chain reaction was set up; the result was that two large pieces of shrapnel passed through the aircraft via the bomb doors. One piece went straight through the aircraft, the only damage being the two holes it made. But the other went through the aircraft floor, damaging the starboard engine's controls, before hitting the flight engineer, Lou Penn. The two starboard engines had to be shut down due to the violent vibration.

So with two engines shut down, it took some time for the aircraft to get back to the airfield, land, and get the engineer out and on his way to hospital, which was down in Nairobi. Sad to say, he didn't make it. He was buried with full military honours in Nairobi Military Cemetery.

The other aircraft used out there was the Twin Pioneer. We used that aircraft as a flying loudspeaker; the idea was to fly up to where the Mau Mau were reckoned to be and broadcast a message to them saying, "Come out of the jungle and you will be all right." How many came out, and if the ones that did come out were indeed all right, we never knew.

Another oddball thing was that we would take a guy up in the Lincolns for a look around and he would map out the area we had bombed. Then on returning to camp, he would set out into the jungle armed to the teeth. I guess he was some sort of mercenary working for the Government.

One day we had the good news that the Army had captured the leader of the Mau Mau; he was Jomo Kenyatta. It was strange to see him as a world statesman when it was all over. Once again we live in a strange old world!

214 SQUADRON IN KENYA: 1954

The Government mercenary, Kenya, 1954

During our stay in Kenya we did have a shopping trip to Aden. It was used as a navigation training flight and we went along to keep the aircraft serviceable. Aden then was a duty-free port and was a very good place to buy what you couldn't get in England at that time. That was before all the troubles out there, and the liners would pick up fuel on the way to Australia, but a few days were long enough to spend there.

It may be of interest to anyone reading this who served in Kenya during the Emergency to hear about the food at Eastleigh. It got so bad that during the AOC's (Air Office Commanding) visit it was decided that no one would go to the midday meal. We stocked up with food from the NAAFI. I don't know who was the engineer of this plot, but all the sections agreed. What made it worse was that during the night someone painted on the mess hall wall "WE WANT FOOD FIT TO EAT". You can guess the reaction to that. With 214 Squadron being temporary men, there were suspects, and I remember one night getting quite a grilling from the RAF police.

FROM PLOUGH TO PLANE

RAF Eastleigh, Kenya, 1954: the writing on the wall

214 SQUADRON IN KENYA: 1954

214 Squadron, Kenya, 1954; Peter is in the 3rd row, 8th from right

We came back to Upwood after about nine months; it was then that we were disbanded and we all went our own ways to other stations and squadrons. I don't think the Lincolns saw any action after that.

So that was end of that story. Kenya became independent, Jomo Kenyatta became the leader, and the rest is history.

After I returned to Upwood from Kenya, and 214 Squadron was disbanded, for the last few months of my five-year engagement I was once again back in the hangar on ASF making major inspections on the remaining Lincoln aircraft, and so ended my Air Force days. I should have done four years in the reserves, but after a year or so they decided to cancel that and stop paying me the few pounds a year I received for being in the reserves - not quite fair as I had fulfilled my five years' side of the deal.

When I was still on the reserves, the Suez Canal crisis blew up. The Egyptian government seized control of the canal, which didn't please the British government. However, as I only had training on the old Lincoln aircraft, I was not much use to them. They did send me the travel documents to go out there, but at the last minute it was cancelled.

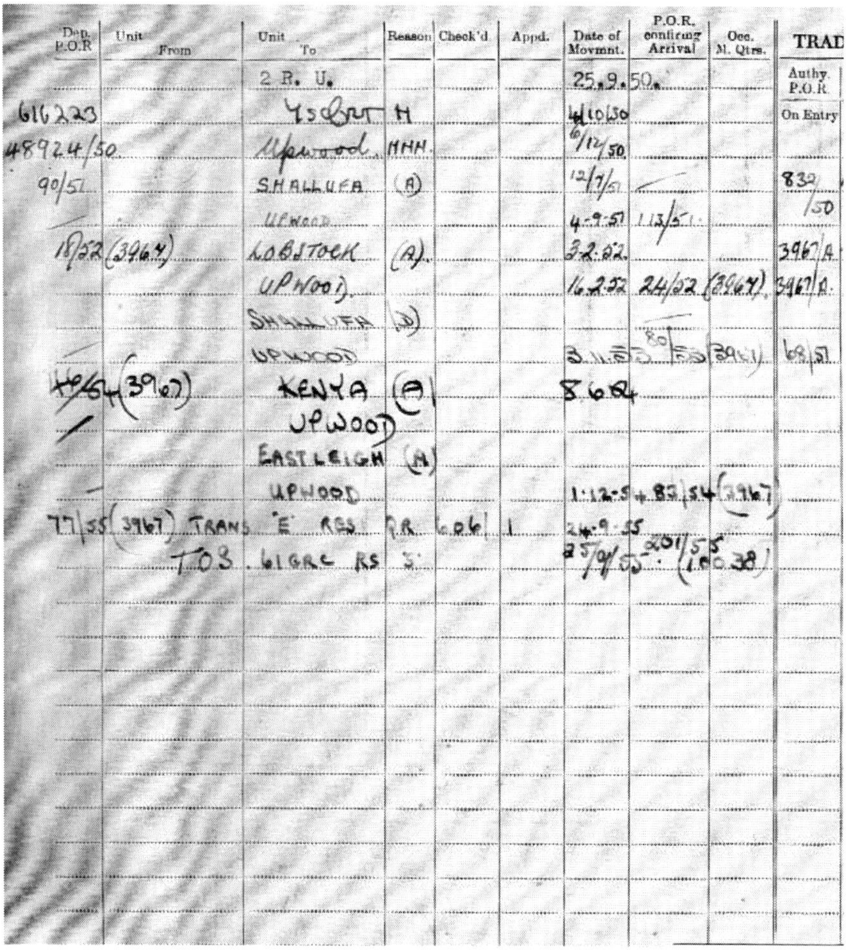

Dep. P.O.R	Unit From	Unit To	Reason	Check'd	Appd.	Date of Movmnt.	P.O.R. confirming Arrival	Occ. M. Qtrs.	TRAD
		2 R. U.				25.9.50.			Authy. P.O.R.
616223		[illegible]	H			4/10/50			On Entry
48924/50		Upwood	HHH			6/12/50			
90/51		SHALLUFA	(A)			12/7/51			832/50
		UPWOOD				4-9-51	113/51		
18/52 (3964)		LOBSTOCK	(A)			3.2.52			3967/A
		UPWOOD				16.2.52	24/52 (3964)		3967/A
		SHALLUFA	(D)						
		UPWOOD				3.11.53	80/53 (3961)		68/51
40/54 (3967)		KENYA	(A)			8.6.54			
		UPWOOD							
		EASTLEIGH	(A)						
		UPWOOD				1.12.54	83/54 (3967)		
77/55 (3967)	TRANS	'E' RES	R.R.	606	1	24.9.55			
	T.O.3	61 GRC RS	S			25/9/55	201/55 (1 & 0.38)		

RAF service record, 1950-55

Civvy Street and
Aviation Traders

Civvy Street and Aviation Traders

My last act in the Air Force was going to Woking for my Demob suit. You could just take the money, but you got out a day early if you went to buy the suit, and that's what I did.

So I found myself back at Market Weston five years after catching the train to Norwich. I couldn't go back on the farm, so once again I packed my bags and left. This time I went to Cambridge for a job at Marshalls of Cambridge, the big aircraft maintenance company.

Marshalls was much the same as the Air Force. They did and still do a lot of work for the Ministry of Defence. The pay was at that time (1955) four shillings an hour, 20p in today's money. We worked 46 hours a week, but we all did a fair amount of overtime. I worked on a Bristol Brigand, Vickers Viscounts and Comet 11s.

I stayed there for about a year and then moved on in 1956 to a Freddie Laker company at Southend airport, Aviation Traders. The only job I could get there was as a detail and assembly fitter on a prototype aircraft which was named the 'Accountant'. This was a Rolls-Royce Dart-powered aircraft, a replacement for the DC3. Looking back, I wasn't really qualified for such work, but when you are young and want a job, it's a case of daring to go 'where angels fear to tread' and all that.

On the first morning I presented myself to the shop foreman, Jim French, an ex-boat builder (as I discovered, boat building and aircraft manufacturing have a lot in common during the first stages of laying out the design). Jim showed me to my bench, right next to the clocking-in clock, so that was a good start - I saw every minute click by throughout the day. To get me started, Jim gave me a drawing and a requisition for the metal that I was to make a part from. I duly got the metal from the stores, went back to my bench

FROM PLOUGH TO PLANE

The 'Accountant'; the aircraft Peter built at Southend airport for Sir Freddie Laker

Building the 'Accountant' at Aviation Traders,
Southend airport, 1956 - a Freddie Laker project (one of these did
fly in 1957, but the project was then cancelled)

A short-nosed Bristol 170 MK 31

A Bristol B170 Mark 32 ready to load at Southend airport, 1958

A Bristol Freighter B170 MK 22 in Sabena livery, Southend airport, 1959

CIVVY STREET AND AVIATION TRADERS

Setting off from Southend to Australia, 1958, working for Air Charter (a Freddie Laker company)

A DC4 G-APNH, Southend airport, 1963 (it later became a Carvair ATL98): "I went to the USA in this aircraft in January 1965. Later on, in Le Touquet airport, France, I led a team of engineers to dismantle and scrap the aircraft after the nose landing gear collapsed on landing."

A Bristol Freighter airborne; "Yes, they did fly!"

The Minor Maintenance Team on the DC3 at Aviation Traders, 1964; Peter is second from right on the second row

and thought, "Now what??" I studied the drawing and decided to make a start. First I had to cut the metal to the right size, and then bend it into shape. The tolerances the drawing gave were very close - only a few thousandths of a inch - so there was no room for error. I took my piece of metal to the guillotine and bending equipment and was standing there looking at what to do, when a old hand in the trade came up to me. Saying, "You alright mate?" he took the metal, cut and bent it, and that was that. All I had to do now was go back to my bench, do the final shaping, and clean the part up ready for the inspection in the view-bay.

I kept my eye on the inspection bench. The system was that if the item passed the inspection, it was then routed to the spray shop for painting. If it failed, then it was placed on the reject bench with a red tag. Fortunately my part didn't come back to the reject bench, so I was happy to get the first job out of the way, albeit with a little help; but we all need that some time.

Over the next few months I became quite skilled at this detail work and asked the foreman for more taxing jobs, as I was enjoying the work and the time clock was not a problem any more. After a while I was moved into the assembly team on the central section of the plane. The other sections were: wings, tail, landing gear, engines, and fuselage. Each section had a darts team; we played in the tea break and lunch hour, and it was very competitive and good fun. Factory work in those days was very different; most probably Heath and Safety would have something to say about darts flying about in the work place if that happened now.

We built one aircraft which flew in the summer of 1957 and went to the Farnborough Air Show that year. We were halfway through the second aircraft when Laker was unable to get the backing to produce it. So overnight the project was cancelled. Laker at that time had acquired all the 252 Percival Prentice planes (an RAF trainer) from the Air Force and had a scheme to convert them to a seven-seater civil aircraft. I don't suppose we had converted any more than about fifteen before that project was shut down too.

When the 'Accountant' project collapsed, most of the staff were out of a job. I stayed on for a while with a few other men. All our time-expired and redundant aircraft, including the 'Accountant', finished up in 'the pot', which was an oil-fired smelting furnace.

Here aircraft were broken up and melted down into ingots. Over the years, many aircraft finished up in the pot at Southend airport. We would be sent over to the boundary, to a site over the other side of the airfield, where the breaking up was done, not only of complete aircraft, but also of brand-new Rolls Royce Merlin engines. Working on a bonus we would dismantle the engines, along with new propellers still in their boxes. What would they be worth in today's market? Some of the aircraft that found their way into the pot were Avro Tudors, Handley Page Hermes, Miles Mathon and well over 200 Percival Prentices, to mention just a few (I could go on listing many more). After spending most of our time repairing aircraft, it was good fun to break them up, but if they had saved some, we would be much richer today in examples of our history of old types of plane.

Eventually the factory shut down completely. That was before any Government redundancy schemes were in operation. I had married Vera by this time and we were living in South Benfleet. We had just bought a bungalow, so it was a bad time to be out of work.

However, during my time with the company I had met up with a mate I had been with in the Air Force. He was working on the other side of Aviation Traders, as they had an airline called Air Charter and a maintenance unit, ATEL, 'Aviation Traders Engineering Ltd'. So after the factory shut I got a job there on the maintenance side.

At that time we had three DC4s and nine Bristol Freight B170s. The DC4s were on trooping contracts and the B170s were on what was called the Channel Air Bridge. That was the car ferry service to the continent carrying people going over on holiday. On the Bristol Freighter or B170 we carried five cars and 23 passengers. The three destinations were Calais, Ostend and Rotterdam. On some days in the height of the holiday season we would fly some 60 movements a day - a very hectic time. I progressed up the ladder over the years I was with ATEL and during some of that time I was a shift leader on the Channel Air Bridge.

As soon as the sea ferries were up and running, the air ferries slowed down, and after a few years they became a thing of the past. The company did try going deeper into Europe with the Carvair (that was how the modified DC4 got its name) to Basle, Geneva,

Strasbourg and Genoa, but that also folded after a short time; the car ferries had finally had their day, and the sea ferries had won.

My first away job as a flight mechanic was with a Bristol Freighter B170 carrying cars to Gothenburg. I then did many flights carrying racehorses around Europe to any airport near a racecourse, such as Paris, Dublin, and several in England. That was quite interesting, but very hard work, as we had to carry the ramp with us to unload the horses. The ramp had to be built up and attached to the sill of the fuselage after the two big nose doors were opened. In addition we took the grooms with us to look after the horses, plus a humane killer gun they could use if one of the horses went wild and the aircraft was in danger of being badly damaged. However, on all the flights I made, the horses were well controlled by the grooms.

Later on in my career at ATEL the DC4s were on a contract to the Ministry of Defence and all three were on the Australia run down to Adelaide, taking a rocket (a military weapon) to be launched from the Wonega test site. I was on the flights to Adelaide as a 'flying spanner' and security guard; because the cargo was highly secret one of us had to sleep on board. There were two flying spanners on the crew, which at that time was made up of two pilots, sometimes three, a navigator and a radio operator. (In my next chapter I will tell you about one particular adventure I had on that route.)

The flight to Australia could take up to five weeks. On the first one I did, the system was to get the aircraft from Southend and back as quickly as possible. We used to have crew stationed down the route, so that when the aircraft landed and the crew had to take a rest, there was a new rested crew ready to continue the flight. We didn't have a crew at each stop, only at two places, first Aden and then Perth.

We would take off from Southend, make our first stop in North Africa at a place then called Castle Benito in Libya, a night stop, and then the next leg was to Wadi Halfa in the Sudan. It was later flooded to make the Aswan Dam. Then we flew to Aden, changed crews, and went on to Karachi in West Pakistan, refuelled, and then on to Colombo in Sri Lanka (Ceylon at that time). From here we flew on to the Cocos Islands in the middle of the Indian Ocean, one of the most relaxing places I have ever visited, with palm trees, sandy beaches, no roads, but only a landing strip - a place I shall never forget. Our next stop was Perth for a crew change, and then we went on to Adelaide.

On the way home there was no load from Adelaide. We would go to Melbourne to pick up some target aircraft called Gindervics. The only difference in our route was that from Ceylon we would fly straight to Aden as we didn't have so much cargo; less cargo meant more fuel, so we could fly further. We still made a stop at the Cocos Islands for fuel, and there we would always load the lower hold with coconuts still in their husks; until the night that one crew chopped down an entire tree and that was the end of the coconuts. By that time however, all the schools around Southend had a coconut in its husk as an exhibit on their nature tables.

I continued to work at Southend with Aviation Traders. In 1967, I obtained my aircraft licence, which was a big step up in my work as an aircraft engineer. I could now certify aircraft maintenance and sign aircraft fit to fly. With this came more money and promotion to inspector/supervisor and then to manager and chief engineer.

By this time aircraft technology had advanced from the piston engine to the jet engine. The Bristol Britannia, Vickers Vanguard, the Canadian CL44, Boeing 707 and the Short's Belfast were but a few that I worked on and certified.

In 1974 Aviation Traders closed the Southend maintenance hangar and we were all made redundant. I had been there some 18 years but not the 21 years that qualified you for the gold watch award - missed out again! Nevertheless I had come a long way in the aircraft world by that time and, as the saying goes: 'As one door closes, another door opens.' In fact another door did open and off I went through it. Fortunately jobs were not too hard to find for a qualified engineer.

After being made redundant from Aviation Traders I got a job at Luton airport with Monarch Airlines. They had the Boeing 720, the Bristol Britannia and the BAC1-11, plus they did the maintenance for Invicta Airlines from Manston on the Vanguards, and as we had done Invicta maintenance at Southend, and I was qualified on that aircraft, I got the job. At that time a big airline at Luton, Court Line, had gone out of business, so there were plenty of aircraft engineers on the market at Luton, but no Vanguard people.

Because I had been made redundant, and Monarch could not fill the position with a local engineer, I was granted Government aid in moving jobs. We moved from Benfleet, where we had lived for eighteen years, to Royston, not that near to Luton, but near enough to be granted moving and legal fees from the redundancy package.

During my time at Monarch I did the Boeing 720 engine and airframe course, so that was a further qualification. Before moving to Royston, while we were still living at Benfleet, I stayed with my sister in Welwyn Garden City during the week, and came home at the weekend.

We eventually moved to Royston, but after a year I left Monarch and got a job with Cargolux in Luxembourg. We couldn't sell our house in Royston, so we shut it up, loaded our car, a Hillman Minx estate, and set off for Luxembourg. I had worked on the Cargolux aircraft at Southend and was well known to the Cargolux engineering management. However it didn't really work out, as accommodation was hard to find. Although Cargolux had promised to help in finding somewhere to live, they were not much help. So after a couple of weeks we decided to come home. I was intending to go back and leave the family at Royston until I could sort something out, but once home I decided to stay put. It was a pity it didn't work out, as we had spent a lot of time organising everything and going to the Luxembourg Embassy, and apart from the accommodation we had most things sorted out.

My next job was with a company at Stansted Airport, Transmeridian Air Cargo. It was an all freight-company and operated CL44 aircraft. Once again, I had worked on these aircraft at Southend, and several of the engineers there had gone straight to Transmeridian when we were all made redundant, so it was not like going into the unknown.

Then Transmeridian was taken over by a Gatwick cargo company and we became British Cargo Airlines. However, that didn't last too long, and once again I was redundant.

Now during this time the Royal Air Force was getting rid of the Short's Belfast aircraft, a large cargo plane. The RAF had ten of this type but because of defence cuts by the Government they were all put up for sale. A company named Pan African ordered all ten but in the event couldn't raise the money to pay for them. So the bank that had agreed to loan Pan African the cost of three of the planes approached what was still then Transmeridian, and went into partnership with the company to try to get them on the civil register. However, during this time Transmeridian had been taken over by British Cargo and so a small company called HeavyLift Cargo Airlines was formed out of the ashes of Transmeridian.

By this time Shorts of Belfast were no longer interested in the Belfast aircraft, so Marshalls of Cambridge took over the design authority of the aircraft.

We moved back to Southend airport, opened up the hangar I had been made redundant from in 1974, and began the task of getting the Belfast on the civil register. They had only been certified for military service at build, and the Civil Aviation Authority required many modifications to bring them up to the required standard for civil use.

HeavyLift Belfast aircraft G-BEPS at Stansted

Carvair to Australia - Nearly

Carvair to Australia - Nearly

In 1965, while I was still working at Southend, I was asked if I would go to Australia with a Carvair as a flight maintenance engineer, or in other words, a 'flying spanner'. Having at that time never been to the USA, I said yes, I would go. The reason for this flight was to take a Black Knight rocket to the testing site in Australia in completed form, so once it was there they could put it straight onto the launching pad and fire it off. Previously we had only taken the rocket in parts that had to be assembled out there before it could be launched.

The aircraft we used was a converted Douglas-built DC4. I had been to Australia a few times with the DC4 taking the rocket parts to Adelaide, and on those flights we were always routed down through Africa, India, the Cocos Islands and into Australia at Perth. This time our route would be via the USA.

The Carvair G-APNH: the aircraft in which the rocket was taken to Australia

The aircraft registration was G-APNH and it had been converted at Southend. It had its air test after completion on the Sunday morning, and in the afternoon that same day we left Southend for RAF Lyneham, so you can see what time we had to sort out any defects found on the air test! When we arrived, the rocket, in a big steel container, was waiting ready for us to load.

The aircraft was loaded, refuelled and inspected ready for an early take-off. All that we had left to do was have a meal, a beer and then of course get off to bed. We went to the bar for a drink, but when it was time for bed, as the representative of the company who had made the rocket tried to get up from his seat, he could hardly move. He was to come with us on the flight, his job being to report back to his company on the conditions the rocket had been subject to during the flight i.e. height, temperatures, pressure and the like. I said to him, "Do you think you should come with us with such a bad back?" His reply was that he had been in the factory all his working life and he wasn't going to miss a trip like this, so that was that and off to bed we all went.

Next morning all was well and after a breakfast in the mess we took off on the next stage of our journey to Keflavik, Iceland. The date was 10th January 1965 - not a good time of the year to be going to that part of the world (in fact rather foolhardy, as you will see later). But we arrived there with only a few small defects to sort out. Now part of the flying spanner's duties was to act as a security man. Seeing this load we had was a secret weapon, one of us had to stay on board each night. On this first night however, the rest of the crew had already gone to the hotel, and by the time we had fixed the snags it was late, so we both decided to stay on board that night.

The next sector of the flight was Keflavik to Duluth, Minnesota, in the USA. We had not been airborne long before we ran into some really heavy snowstorms. Now being an unpressurised aircraft meant we could not go up above the bad weather because of the lack of oxygen. I will explain the de-icing system on our aircraft. The wings and tail were de-iced by having rubber boots on the leading edges; when ice formed the boot would inflate and the ice would break away. That was all right as long as the engines were running and producing the air for that system. The propellers on the other hand were a different ball game. That system used a de-icing fluid that was pumped to the props, and with the rotation of the props was slung over them, so that ice didn't form. The only trouble with this system was that we only had so much fluid in the tank and we were using that up fast. The only solution was to climb out of the bad weather and take a chance with the lack of air.

After we had climbed out of the bad weather, all went well and we arrived at Duluth all in one piece; the aircraft was serviceable but it was cold. We had thought Iceland was cold but in Minnesota the temperature was -34°, and it was my turn to stay with the plane. It was far too cold to sleep on board, so I stayed the night in the terminal building where I could see our plane. It was not the way the security people would have wanted it done but I wasn't going to freeze to death, and if someone wanted to blow the rocket up, I was in the best place.

In the morning the plane was really iced up so before we could go anywhere we had to de-ice it. This meant getting heaters to blow hot air into the front and back of the engines before we could even attempt to start them. Eventually we were ready to start and off we went; next stop California and very glad we were to be getting out of that cold, anticipating that San Francisco had to be warmer, even at this time of the year. Sure enough it was a lovely warm night when we landed.

Tonight was my first night in a hotel and a real bed since I had left home. I stayed and helped Frank, the other engineer, with today's defects and then made my way to the hotel, but it was not going to be as simple as I thought. I went to the handling agent, which was Qantas Airlines of Australia, and they gave me a taxi ticket to get from the airport to the Drake Wilkshire hotel in town. "Just take any yellow cab outside," they said, and that is what I did.

"Drake Wilkshire please," I said and got in the taxi.

"First time in the US buddy?" asked the cab driver.

"No," I said. I did not want to appear to be a 'green horn'.

"You from Australia buddy?"

"No, England," was my reply. He drove me around pointing out all the landmarks but all I wanted was to get to the hotel, have a hot bath, a meal and my bed.

We arrived at the hotel at last. During the ride I hadn't taken much notice of the meter as I had a ticket that would take care of the fare. But no! When I gave him the ticket he went mad because what he had done was to take a very long way round for a big fare. He called his office on the radio and then took me to the taxi office. I wasn't going to pay any more, so in the end he had to take me back to the hotel - not the happiest taxi driver in the US that night. For my part I was only glad to get to the hotel.

I checked in and the hotel porter would insist he carried my very small bag to my room. I said, "No I can manage." But he wouldn't let me. Once inside the room he showed me how to turn on the light, switch on the TV, run the bath and draw the curtains. Then I fell in. He was after his tip. I only had a few coins in my pocket so I gave him them. He looked at them and then said, "Hell, man! These are worth nothing."

"That's all I have. Goodnight and thanks," I said, making the porter the second unhappy man of the night. It had been a very long day and all I cared about right now was getting to bed.

In the morning the weather was good - warm and sunny - but the winds for the next stage were too high and we wouldn't be going anywhere today. The company rep with the bad back was no better, so as we had a day with no flying, he thought he might as well see a doctor about his back, and that was the last we saw of him. I think he must have been in a great deal of pain to have gone to the doctor at all.

We went up to the airport and seeing what a nice day it was, we decided to open all the doors and let the wind blow through the aircraft to dry it out. Because of the big load we were carrying, the aircraft had been stripped of all the unnecessary equipment. That included the hydraulic system to open the big nose door, so we had to pull it open with ropes, using one to pull it open and the other to restrain it from opening too fast.

Then we waited for a few days to see what was happening about the man with the bad back. It turned out that he was to stay in hospital and a new company rep would come out to take his place. When the new rep arrived, we began to prepare for the next stage.

By now the weather was good; the winds had moved from head to tail, so we decided to go. We did all the inspections on the plane, full ground runs, then refuelled after the engine checks so we had as much fuel on board as we could get. This was one of the overweight take-offs we had got permission for from the civil aviation authority, because with the normal maximum take-off weight we could not have made it to the next stop, that being Honolulu. Off we set, all going well, when we received a radio message telling us to return to San Francisco. We found out later that the data the company had received had been analysed, and they had found that the solid fuel in the rocket would be cracked due to the very low temperature it had been subject to.

So round we turned and headed back. Now we still had a lot of fuel on board - too much to land with, so we had to dump some to bring the landing weight down to the correct permissible weight. However one of the dump chutes stuck open and because it was still dumping fuel on landing, we had a fire engine following us back to the parking bay. So there we were, right back where we had started from. What next?

After a night in the hotel all that was left was to take the rocket back to England. We took the southern route home - that was via San Francisco, Dallas, Bermuda, the Azores, Lyneham and Southend. Why we never went that way outbound I never did know. It would have been much better for everyone, including the rocket, and maybe the solid fuel would not have cracked and we would have been the first Carvair into Australia, as well as the first Carvair into the USA.

Eventually we arrived back at Lyneham, offloaded, de-briefed with the security men, and it was home to Southend and back to work in the hangar. Another saga of my aviation life over ...

Engine Changes over the Years

Engine Changes over the Years

During my time working for airlines over some fifty years, I have changed many engines on many types of aircraft and in many places around the world. My first was on a DC4 in Berlin, operated by Air Charter, a Freddie Laker airline, which was in 1959.

During my time as Line Maintenance Manager for HeavyLift Cargo Airlines at Stansted, I was always on call. If one of their aircraft had any defects that they required help with, they would call me direct from the aircraft, either airborne or from the ground, from any part of the world. It seemed to happen mostly in the middle of the night when I was in a deep sleep. The aircraft could be en route, say, from Singapore going to Australia. They would call me at home, and then I would try to advise them on the course of action they should take, telling them to call me back once they had landed.

My last engine change was in Cape Canaveral on a Short's Belfast operated by HeavyLift Cargo Airlines, which was in 1998, so that made some forty years tramping around the world fixing broken down aircraft.

My First Engine Change in Berlin

One day I was working at Southend when we got a call from one of our DC4s operating out of Berlin that one of their engines had failed. The call came late one afternoon and the company wanted the aircraft back in service the next morning. This was a tall order, plus the fact that the only engine we had in the engine bay was built as a left-hand installation and, yes, the failed engine was on the right. The only part that 'handed' the engine was the exhaust system, but we didn't have time to change over the exhaust and still make the deadline. So we loaded the engine, plus the right-hand exhaust system, and off we went, en route to Berlin. Our journey was to take us about four hours, so on the way the other engineer on board and I changed the exhaust over to the correct hand. Thus when we landed at Berlin, the engine was ready to fit.

Once we were in the hangar at Tempelhof, we started to change the engine. By the morning we had the engine changed, had done all the ground checks, run the engine, and it was on the ramp ready for the first service of the day. Then we had a wash and a meal and flew back to Southend - all within 24 hours. So you can guess that we never saw much of Berlin, not from the ground at least, but as I have said before, that was the way it was.

The Engine Change in Bangkok

When I was still working at Aviation Traders at Southend I did an airframe course on the CL44 aircraft. One day during the course a knock came on the classroom door; it was the hangar foreman. "Could I have a word with Peter?" he asked the instructor. Once outside the classroom he said, "Will you do me a big favour? I need someone to go to Bangkok to change an engine on a CL44." I had a licence on that engine, so I was well qualified to carry out the job. I gave it some thought and said, "Tell you what I'll do - if I can take two engineers of my choice, I will go." And he readily agreed to that, so it was all systems go.

The next move was to get all the equipment together, engine change kit, manuals and seals etc. We went home to pack an overnight kit as we thought three days at the most would do it - one to get there, one to change the engine and one to get home. How wrong can you be!

The aircraft that would take us to Bangkok was at Gatwick but the spare engine was at Frankfurt. So that was our first stop. We arrived, loaded the engine, a fourth engineer joined us there, we refuelled, and set off on our way to Beirut. That was to be just a refuelling stop as well as being the home of the aircraft.

I will now give you the background to the operation. The aircraft was owned by an American company called Seaboardworld, operated by the Lebanese airline TMA. The aircrew were from a Gatwick airline called Tradewinds. We, the engine change crew,

ENGINE CHANGES OVER THE YEARS

worked at Aviation Traders of Southend. You will see later on how this set-up wasn't very helpful to us.

We left Beirut after refuelling with enough fuel to make Bangkok - a long flight and not all that comfortable, our plane being a cargo aircraft. After about ten hours we landed in Bangkok in the early hours and decided we should have a rest before starting work. So it was down to the hotel, a meal and into bed. It had been a long time since that knock came on the classroom door.

We had booked an early call and the phone rang after what seemed only a short while since falling asleep. So down to breakfast and up to the airport ready for the engine change, which went very well. By the end of the day we had fitted the new engine, done all the checks, made full power runs, and declared the aircraft fully serviceable.

We tidied up, put all the equipment away and had a wash with the aid of a hosepipe in the corner of the hangar. I think it was used for washing down, but it was cool and I am sure we were somewhat cleaner.

Then over to the terminal building for a meal and to buy a few presents to take home. We changed some money into local currency, only to find there was very little to buy. When we tried to change our local money back, they didn't want to know. So we were left with a pocket of all but useless notes.

Once the company knew the aircraft was serviceable, they wanted it back in the air earning money. The flight crew was soon in the aircraft and ready to go, so we set off for Bombay. The new engine was going well and we were pleased about that, but then it all went wrong in a big way.

On taxiing in at Bombay, standing on the tarmac was a CL44 of the same company, and (would you believe it?) that also wanted an engine change. At first we didn't think it would involve us, as there was no engine for us to change, but once again how wrong can you be? The airline had made the decision to offload the unserviceable aircraft, put the load onto our plane, and ferry the other one back to Stansted on three engines. Now this was our problem: with one plane on its way to Japan (and it's law that only the operating crew can be on board on a three-engine ferry) we were left standing on the tarmac with no plane.

We found a hotel, or actually a rundown hut, near the beach, but then it was to be only for one night, or so we thought. As soon as it was light we packed our bags and went up to the airport, straight to the Air India desk; being airline staff we thought we should have no problem with a ticket back to London. At first all went fine, and they checked our bags in. All we had to do was to wait for someone to give the authority for the issue of the tickets. We waited and waited but he never came; so no ticket, no flight. They offloaded our bags, and it was back to the 'Arupan' (that was the name of the shack on the beach).

Next morning we had to try and see what we could do about getting home. It was nigh on impossible to phone the company and we had very little money - there were no plastic cards at that time. One of the engineers went down to the British Embassy, but they were not much help. They told us to come back in a week if we got no help from our company. After a day or so we met the crew of another CL44 going south. They lent us some money and told us they should be back through Bombay in a few days, so that was good news.

We were up at the airport every day waiting for that CL44 to land. On the third day, out of the blue, came a lovely sight - a TMA CL44. We met the aircraft, had a word with the crew, and they were happy to take us with them. Only one thing worried me and that was that there seemed to be more people wanting to fly on the plane then there were seats available for them. So I said to the other two, "Let's get on board, and don't get out of the seats until we are in the air." And that is what we did. The company wanted to offload us and make seats available to some company employees, but we stayed put.

At last we were in the air and heading in the right direction, but once again more trouble lay ahead. The company had omitted, before leaving India, to get over-flight clearance for Pakistan. So when we reached the Pakistan border, the Pakistan air traffic control requested that we land, and not continue to fly over their airspace. If we failed to land, they would send up their fighters to escort us down. I think the crew were in a bit of panic, so they turned round, headed back to Bombay and landed there. Once more we were back down to the Arupan, our second home, or so it felt.

ENGINE CHANGES OVER THE YEARS

They couldn't get clearance to fly over Pakistan next, so we had to make a landing there, and a night stop. We did some shopping while we were there; I bought a camel stool and a few other things.

Well, so far, so good; we were halfway home, with the next stop Beirut. We got there without further ado and landed safely in Beirut. We were met by the engineer we had picked up in Frankfurt. Next it was down to a very good hotel, a nice meal, a night's rest, and in the morning a flight on a Boeing 707 back to London.

At London airport we did a deal with a taxi driver to take us three to Southend for £10. Little did he know, the plan was to drop me off at Benfleet, George at Eastwood, and lastly John at Rochford. I don't think he bargained for all that running around, but we were home and that was all that mattered. Not such a simple engine change!

DC 8, Stansted, 1979; the aircraft Peter worked on during an engine change in Hong Kong, 1980

CL44 G-AZIN, Stansted, 1979; the aircraft Peter worked on during an engine change in Bombay, India, 1981

DC 8-54 G-BTAC, Stansted, 1979; the aircraft Peter worked on during engine changes in Hong Kong, 1979 and in Delhi, 1980

ENGINE CHANGES OVER THE YEARS

The Delhi Adventure

During the time I was working at Southend with ATEL, as I have said, we operated nine Bristol Freighters on the car ferry run over to the continent. Over the years we had many engine failures at Ostend, Calais and Rotterdam; they were easy to contend with as they were so near our main base, and with so many flights a day, it was fairly straightforward.

A much greater problem was when we had a defect further afield; we then had the task of getting the new engine to where the aircraft was stranded. Sometimes that could be some remote airfield on the other side of the world.

First you had to make sure before the engine was despatched that you had everything you needed - not only the engine, but the correct tools, manuals, oil for the new engine, and all the required legal paperwork.

One of the engine changes I did was on a DC8 when I was working for Transmeridian Air Cargo at Stansted. I was attending a Flight Engineers' course on the DC8 aircraft at the KLM training school in Amsterdam, when a call came to the classroom from our Chief Engineer at Stansted, wanting to talk to me (this was just like what happened when I was sent to do the engine change in Bangkok).

"Peter," he said, "We have a engine change on one of the DC8s in India, Delhi. Can you leave the classroom, go to the KLM ticket office, get a flight to Luxembourg and when you get there, go to the Airline Cargolux. They are loaning us a engine and flying it down to Delhi in the morning. So you go with the engine, and I will send you two engineers from Stansted to give you a hand with the change."

I had actually worked for Cargolux for a short time before I joined TMAC. I duly got the flight to Luxembourg, found the engineers at Cargolux, got the engine loaded, had a night in the local hotel, and in the morning took off in one of their DC8s non-stop to Delhi with the engine on board; so far so good.

On arrival at Delhi, the two engineers were waiting for me, so after offloading the engine, we made a start on changing the engine. One of the big problems down the route is always the equipment required to carry out any maintenance on your aircraft, such as stepladders and working stands. On the DC8 it was not as bad as on some aircraft, as the engines are low to the ground. The Belfast on the other hand was a real problem, with the engines being so high up from the ground. We normally tried to beg, borrow or steal the passenger steps, but after we completed the work, there was always the job of cleaning the steps of all the oil that had dripped on them during the work we had carried out. If we didn't do this, a big bill was sent to the company for cleaning the equipment and any oil that had dripped onto the floor. We were also aware that we might have to return one day to fix another aircraft, and it was not a good idea to leave the place in a mess.

Back to the work in hand; we started to change the engine but night-time was approaching. Not only that, but there was a large pack of wild dogs lying under and around the aircraft, so the wise thing to do was to call it a day, go to the hotel and come back in the morning when we could see what we were doing. Maybe the dogs would have drifted to another spot by then. They didn't look too healthy to me, so it was best they didn't get too near us; at least in daylight we could keep a eye on them.

Next day we completed the work, carried out all the checks, did the engine runs, and found all was well, so we packed all the gear away. I signed the aircraft logbook that all the work was completed and that the aircraft was fully serviceable and ready to fly on its way.

We got a flight back to London - another job completed. I never did go back to KLM to complete the course on the DC8. However, I had done the airframe and engine course at Britannia's engineering training school at Luton Airport on the DC8, and had the CAA type rated licence on the aircraft, so missing the end of the course wasn't the end of the world.

The Puffin Club

Yes, I am a fully paid-up member of the 'St John's Newfoundland Puffin Club', so let me explain how that came about.

When I was working for HeavyLift, one particular night I got a call from one of our Belfast aircraft that was on the ground in Newfoundland. It was the captain speaking: "Peter! We've got a big problem here in St John's and we need more than advice - we need you to come out here and see what can be done. We think you'll need to bring a structural engineer with you too, as the plane has been extensively damaged." So as not to put the family to any more disruption at that time of night, I told him that I would go into work and call him from the office.

From vast experience over the years, I knew I needed to pack a bag; I always carried my passport in my briefcase. Then it was off down the M11 motorway again to Stansted. When I got into the office, my boss was already there. He didn't want to send a structural engineer with me; as we didn't have one working with the airline, it would have meant getting someone in from outside (expensive). So I said that I would go and have a look, and if I couldn't sort it out, I would call him and take it from there. Operations called a taxi to take me to Heathrow, and then they sorted out the airline tickets. As they couldn't get a direct flight to St John's, I had to fly to Halifax and then get a flight back to St John's.

On arrival I was met by the crew. I went straight to the aircraft to carry out the initial inspection and get the crew's report as to what had happened to cause the damage.

Belfast aircraft at St John's, Newfoundland, 1998

I found serious damage to the aircraft's tail section, due to an accident when unloading a very heavy generator that had been transported to St John's. Before I go any further, I should describe the Belfast's loading and unloading procedure. The aircraft has a rear ramp, and a large rear door; on loading the aircraft, the ramp is lowered and the door is raised, providing a large aperture through which to load the aircraft. On the first stage of loading, the ramp is raised so that it is in line with the truck, and the load is placed on the ramp ready to be winched into the cargo hold. To prevent the aircraft tipping onto its rear, two hydraulic jacks have to be placed under the rear fuselage.

However, on unloading the generator, the load master forgot to place the ramp support jacks in place. As the heavy generator was being pulled out of the aircraft onto the ramp with no support jacks in place, the aircraft's nose came up and the rear of the aircraft came down. As it dropped, it fell onto the fixed crane on the truck (which was positioned in line with the ramp, so the generator could be pulled straight onto the truck). In doing so, it punched a very large hole in the tail section. Ever after this incident, the load master was known as 'Tippey Walker'. It has to be said that the company was not best pleased with Tippey, but that was the last time he forgot to install the ramp support jacks.

So the damage was very extensive but not beyond my capability. I called Stansted and requested two engineers to assist me with the repairs and the spare parts, and I specified the tools I would require to carry out the repair. I must just point out that this happened in the height of the winter. As we could not get the large Belfast into any hangar and out of the very severe weather, it was going to be another outside job. However, with my many years working outside in all conditions, varying between the heat of the desert and the cold of St John's, it was all part of an aircraft engineer's work.

What's all this to do with the Puffin Club? I had better get round to explaining. After work one night, two of the flight crew and myself decided to go out for a beer. We walked into town, and when we found a bar that looked okay, we went in. But as soon as we entered the bar, a very loud bell rang out, and an equally loud voice also rang out: "Strangers in the bar!" We were all a bit taken aback by this, but still went on up to the bar. The barman asked us,

ENGINE CHANGES OVER THE YEARS

"Are you members of the Puffin Club? You can't drink in here without being members."

"How can we become members?" we asked, as by this time, instead of going to find another bar, we were all rather intrigued to know what it was all about.

"If you care to become a member, you will have to go through a small initiation ceremony," said the barman,

We all agreed, so the barman got three glasses and filled them with 'Screech' (which I will tell you about shortly). He then said, "You must drink it down in one go and not spill a single drop; we will be watching you very closely. If you pass that test, you must then kiss a puffin."

They had a stuffed puffin behind the bar for just such occasions; I won't tell you where we had to kiss it! We all passed with flying colours and were presented with a Diploma, which meant that from then on we were full member of the St John's Puffin Club, and could drink all the beer or 'Screech' we could take. Now I think the cunning plan behind all this was to sell the three of us a tee shirt. But although it turned out to be an expensive tee shirt, it was an experience that made it worth the money.

'Screech' bottle

The 'last act' of that night's entertainment was yet to come. As we all left the bar, I got talking to some of the locals, then turned round to go home, only to find that the rest of the crew had vanished off the face of the earth. Unfortunately, I didn't know the name of the hotel we were staying in. In the old days, you always would have known, as the hotel's name would be on the key tab. But with the new door security systems these days, you only get a plastic card with an electronic chip. So I spent an hour walking the dark streets of St John's, trying to find the way home. I came across it eventually, and was very glad I did, because even with a glass or two of 'Screech' inside me, I was getting very cold.

Now let me tell you about that wonderful drink 'Screech'; here is the tale. Long before the Liquor Board was created to take alcohol under its benevolent wing, Demerara rum was a mainstay of the Newfoundland diet, with salt fish traded to the West Indies in exchanged for rum. When the Government took control of the liquor business in the early twentieth century, it began selling the rum in an unlabelled bottle. The product might have remained permanently nameless, except for the influx of American servicemen to the island during the Second World War.

As the story goes, the Commanding Officer of the original detachment was having his first taste of Newfoundland hospitality and, imitating the custom of his host, downed his drink in one gulp. The American's blood-curdling howl, when he regained his breath after this, brought the sympathetic and curious from miles around, rushing to the house to find out what was going on.

The first to arrive was a garrulous old American sergeant, who pounded on the door and demanded, "What the cripes was that ungodly screech??"

The taciturn Newfoundlander who had answered the door replied simply, "The screech? 'Tis the rum, me son."

Thus was born a legend. As word of the incident spread, the soldiers determined to try this mysterious 'Screech', and finding its effects as devastating as the name implies, adopted it as their favourite.

The opportunistic Liquor Board pounced on the name and reputation, and began labelling the rum as 'Newfoundland Screech',

ENGINE CHANGES OVER THE YEARS

the most popular brand on the island, even today (although now they use Jamaican rum).

What am I doing in St John's? Oh yes, we have a badly damaged aircraft to repair somewhere out there on the ramp! In spite of the cold, snow, wind, and a lack of equipment, we pressed on. Eventually, after a couple of weeks with a drop of 'Screech' after work, the aircraft was repaired. We got it ready to take to the air and wing its way over the Atlantic Ocean to Stansted and home. After about ten hours in the air, we landed at Stansted, with a few bottles of 'Screech' on board to see us safely through the winter.

My Last Engine Change in Cape Canaveral

The last, and one of the most interesting, places I have been to was Cape Canaveral, Florida in the USA, where I worked on a Belfast engine change. This was during my last days as Line Maintenance Manager for Heavylift Cargo Airlines at Stansted. My team had been reduced to seven engineers, and we had two Belfast freighter aircraft to maintain, so I was more than likely to be involved in any maintenance work.

Heavylift had over the years carried many types of cargo, including the third stage for the Arian rocket which was launched from Cayenne, French Guyana. They also took satellites to the launching pads at Cape Canaveral, which is where this engine change story takes place.

The engine fitted to the Belfast aircraft was a Rolls Royce Tyne. Rolls Royce named their engines after rivers, for example, Dart, Avon, Spey, Conway, Trent, and others. On the Tyne, one of the inspections that was carried out every 25 flying hours was to inspect the magnetic plug in the oil system for metal contamination. This was normally done by ground engineers at Stansted, but on long flights it was carried out by the flight engineer, Cape Canaveral being one such flight.

On this occasion, the flight engineer duly carried out the inspection and found metal on the plug. The crew then called me at home, requesting me to go out and inspect the engine to check its serviceability, and whether they could continue with the flight. One added complication was that a Civil Aviation Authority inspector was on board, so everything had to be according to the book, although that was always the case anyway, as flight safety is paramount at all times, regardless of commercial pressure and cost.

So after giving it some thought, I made the decision to change the engine, which was not an easy decision because of the cost. However, if there was enough metal on the plug for the flight engineer to be concerned, and given that I knew the engine well, it would have unprofessional of me to tell the crew to continue the flight without further investigation.

Our next task was to get a serviceable engine on the way. We had done this many times over the years, so it was routine procedure for the company. Once again I was winging my way to an aircraft that needed some help to get it back in the air.

I had taken an engineer with me, so with the help of the aircraft's load master, we had enough personnel to change the engine. When we arrived at the aircraft and started to get organised, we found that the equipment they had there was second to none. There was a great working platform that went under the engine, with plenty of room to move around; it was the best equipment for the job you could get. Laid on top of the platform was a thick layer of a foam-type matting, to catch any oil that would drip out of the engine during the work, and any thing else that would fall. After the work was completed, the whole of the matting was rolled up and taken away to be disposed of. It was the most environmentally friendly engine change ever. We were to learn more about the American environment laws on our way home.

Working at Cape Canaveral, I did see the rocket launching pads, but no actual launch - only the odd crocodile lying around in the storm ditches. That was a bit scary, but they didn't seem to worry about an engine change team from England working on their patch and were content just to watch. I didn't think they were hungry enough to fancy us for dinner.

The engine change went well and the aircraft was back ready to go; we got a good night's rest and the next day we took off for Bangor, Maine. On our way home, on landing at Bangor, as is normal, we refuelled the aircraft with enough fuel to fly the Atlantic Ocean and back to Stansted, which was about a ten-hour flight. There was a lot of fuel on board, as we were to find out in the morning.

ENGINE CHANGES OVER THE YEARS

Next morning we were all out of bed really early; it was still dark and very cold. After a quick breakfast we took a taxi to the airport and went out to the aircraft to get it ready for the long flight back to Stansted, with a fuel load of 80,000 pounds (about 11,500 Imperial gallons). We did all the pre-flight checks and got on board ready to start engines, but when we went to start No 4 engine, no way could we bring it to life, even after trying all the tricks we had learnt over the years. Would it start? NO!

After we had shut down the other three engines and put the thinking cap on, it was a question of "What now?" On the plus side, we had the old engine in the back of the aircraft that we had just changed, so we had plenty of spares to work with. What is more, I had a good idea what the problem could be - the 'high pressure' fuel pump. It was still dark and cold, but we had to make a start on fixing the defect. I said to Mike, the other engineer, "You remove the pump from the old engine and I'll get the load master to help me take the pump off the No 4."

At this stage, let me just explain the Belfast fuel system. The fuel tanks were part of the wing (some of the older aircraft had separate tanks). With this type of structure, the Belfasts were prone to leaking, even as new aircraft. This particular aircraft had done over 20,000 flying hours, which of course made it more likely to leak under the conditions of that morning - the full tanks and the very cold temperature.

There were a few small fuel leaks on the underside of the wings, but that was not a great problem. My big problem was getting that engine to start so we could get on our way. I had just started to remove the pump when a truck came alongside our plane and an airport official got out.

Walking under the wing, he asked, "Who is the Crew Chief with the ship?"

"I am," I said.

"What is it that's dripping from the wing?" he asked.

"Only water," I said, "The warm fuel has melted the thin layer of ice that has formed overnight; we have just refuelled."

That was maybe a little bit of a lie, but I knew the consequences if he was told it was fuel. I knew that the aircraft was safe to fly with a small leak well within maintenance manual requirements, so I pressed on with the job in hand. By this time we had changed the pumps and were about ready to see if the pump change had fixed our problem. Now it was daylight, who should pay us another visit but our friend the airport man from the National Guard. He went under the wing again, but this time he caught a drop of the dripping fluid in his hand, had a sniff and that was that - all hell was let loose.

He immediately called his boss and ordered us to stop work; the next thing we knew we were surrounded by fire trucks and police cars. Then a truck full of sandbags arrived and they built a wall right around the plane. Next, a team of men with what I can only describe as garden rollers, only with a thick foam around the outside of the roller and a metal bar to squeeze the foam. As the roller went over the top, the fluid was then drained into a container hanging inside. There was very little fuel, but lots of water. It was all a little bit over the top I would say, but as we found out during that engine change at Cape Canaveral, the American environmental laws are something else.

What now? With all this emergency activity, we couldn't get onto our aircraft to check out whether the engine would start. I asked the head man if they would allow me to check the engine and see if it would start. After numerous radio calls - I think they contacted the President (not really!) - they did finally allow me on board and, yes, I found that we had fixed the defect. While I was checking out No 4 engine they sent the other engineer to ask me to run all four engines to use up all the fuel on board so we could fix the leaks. Now as I have said, we had enough fuel to get us back to Stansted and it would have taken me all day and more to use all that fuel up; plus you can't run engines on the ground all that time, due to the cooling of the engine. So that was not really on. I shut down the engines and then had a talk with the airport officials to decide what we could do.

The only possible solution was to try and defuel into a fuel truck and then attempt to fix the leaks. I knew it would be well nigh impossible out on the ramp with the weather being so cold and wet. Even in the hangar at Southend, with all the equipment and a warm

ENGINE CHANGES OVER THE YEARS

hangar, it would still be one hell of a job. That said, what chance did I have on a cold apron at Bangor airport? Nevertheless, I had to be seen to be doing what was required to satisfy the American authorities. The fuel company had a empty fuel truck we could use, but it was parked outside and therefore frozen solid, so our first priority was to get it into a warm hangar and thaw it out before we could make a start with the defuelling.

When the truck was thawed out, we offloaded as much fuel as it would take and then applied some external sealant to the leaking areas. It was not that successful, but it looked good to the watching airport authorities. We told them that we would leave the refuelling until the morning, which would give the sealant time to set.

I then said to the pilot, "What we will have to do is go back to the hotel, get up early in the morning, have breakfast and get back up to the airport. We should refuel and get airborne straightaway before the fuel has time to leak and while the airport is still quiet."

So that is what we did; all went well and we were soon on our way home. That was the last time I changed any engine overseas, but I think that by that time I had certainly done my fair share. On the next major inspection of that aircraft, all the fuel leaks were seen to be fixed and the old Belfast was good for a few more years.

After a year or so, HeavyLift Cargo Airlines went out of business and the two Belfasts were sold to some company in Australia. That was the end of twenty years of good and interesting work on that great aircraft, and it rounds off my many tales about fixing broken down aircraft.

FROM PLOUGH TO PLANE

At the Paris Air Show, 1983

On holiday in Brussels with Vera, 1984

ENGINE CHANGES OVER THE YEARS

Mealtime during an air show in Czech Republic, 1998

My Years with the 'Sally B'

My Years with the 'Sally B'

I first saw the B17s on 23rd June 1943 when 17 of them landed at Knettishall air base in Suffolk, Knettishall being the next village to Market Weston, where I lived. They were the first aircraft to form the 388th Bomb Group.

The B17s were part of the many aircraft that filled the skies over East Anglia during the Second World War and the two years that the 388th Bomb Group was based at Knettishall. The first of these planes to leave for the USA after the war in Europe had ended was called 'Five Grand' - that being the 5000th B17 built since Pearl Harbor, and it was signed by all the Boeing people involved in building it. During the time Five Grand was on operations with the 96th Bomb Group, it carried out 78 bombing missions, two food-dropping flights and two POW flights bringing prisoners back to England.

That was the beginning of my association with the famous B17. At that time I was just 12 years old, and never did I think for one moment that some forty years later I would be involved with one of the 12,731 B17s built by Boeing, Douglas and Lockheed. However, in 1984 I became a major part of the team responsible for the maintenance of one such aircraft, the Boeing B17G s/n 44-85784, better known as Sally B, built by Lockheed in 1944.

There has been much written about the history of Sally B, so I will just tell you my story and about my involvement with the only flying B17 in England. The plane is based at the Imperial War Museum, Duxford, Cambridge (but is not part of the Museum's stock).

At this time I wasn't that concerned at all about air shows; I think I had only been to one or two, one being at RAF Mildenhall, and the other may have been on Southend seafront, as we lived in Southend for many years. Since I was earning my living maintaining aircraft I didn't think paying to watch planes fly was a ideal way of spending my free time. I was more interested in antique clocks and medal collecting, which both went by the board once Sally B came into my life.

When I was working for Aviation Traders I had seen Sally B pass through Southend airport on its way to Europe to do the air shows over there. At that time of day you had to clear customs before leaving the UK. Now we can do that from Duxford

So how did I get involved? At that time I was working for a cargo airline at Stansted and one of our captains was Keith Sissons. Keith was also at that time the Sally B's Chief Pilot. I had worked with Keith over the years with different airlines that flew big piston aircraft such as the DC3s, DC4s, DC6s, and DC7s, plus the sleeve-valve-engined Bristol freighter, so he knew I was a big piston engine man.

During one of my engine changes down the route to Delhi, India, on one of the company's Belfast aircraft, the captain was Keith. After work one night I was having a beer with him and talking over the old days, when he said, "We could do with you on the B17 - how about it?" After another beer or two I said I would think it over, and then promptly forgot all about it.

A little while later, I was having a rare day off and mowing the lawn when the telephone rang. Typical, I thought, take one day off work and they want me to come in and sort out a defect, change an engine or, worse still, pack a bag and go off down the route. But no, it was a lady's voice on the other end, not English but Scandinavian-sounding.

She said, "Peter Brown?"

"Yes, that's me," I replied.

"Keith has given me your name and number and said you would help us out," she said, "This is Elly Sallingboe, the operator of Sally B. My engineers have just fitted a new engine and they can't get it started. Can you help? I have a engineer at Duxford to help you."

Leaving a note for my wife, who had gone out, saying "Gone to Duxford", off I set. On arriving at Duxford, there was Sally B standing on the airfield in the cold and windy weather. "What am I doing here?" I said to myself. That is a question I have asked myself frequently over the years in many corners of the world. Maybe I have engine oil in my veins!

Back to the problem on hand: the engine wouldn't start. I had not worked on piston aircraft for many years as I had moved into the jet age but, as they say, it's like riding a bike - you don't forget. I didn't actually fix the defect but found the reason the engine refused to start. When the engineer returned from a training flight he had been taking in the Tiger Moth, I was able to tell him what to do, and that was the start of many years as Chief Engineer with Sally B, as part of the B17 Preservation.

Over the years I have been to many air shows in many places around the world: Sweden, France, Belgium, Poland, Ireland, Denmark, Italy, the Channel Islands, the Isle of Man, Czech Republic, The Netherlands, Scotland, Wales and Switzerland. There are far too many to tell you about them all, so I will just write about the few that were really the best. In the next chapter I will recount my part in the making of the film 'The Memphis Belle', featuring our B17.

The Longest Flight

The longest flight I have done in Sally B was to Forli in Northern Italy, which took seven hours and five minutes - a long way in a B17, but a great and very interesting flight.

We had full tanks of fuel - some 1700 US gallons of gas 100LL - and a take-off weight of 47,074 pounds. Given that the maximum take-off weight for the aircraft is 50,000 pounds, you can see we still had some to spare. After all the engines were started and all the checks were carried out, and with everything looking fine, it was chocks away. At 08.05 hours we were on our way, taking off from Duxford at 08.15 hours.

We then crossed the Thames at 08.35 hours and the French coast at 09.15 hours - so far so good, and we settled down for a very long flight. Our route took us over France to the south coast and along the French Riviera to Genoa, crossing the French/Italian border at Savona. Along the French coast we had to come down to 500 feet, giving us a brilliant view of Monaco and all the small towns and villages spread along the coastline. After Genoa, we headed for Bologna, passing over the Imolfi Race Track, and reached Forli some seven hours, five minutes after leaving Duxford.

The reason we had to fly the long way round was because of bad weather over the Alps. As you will see from the map, flying over the Alps would have been much shorter and of course quicker. We came home over the Alps as the weather had improved and that made our flight over a hour quicker, so it was a very interesting flight both ways.

We were greeted at Forli with great enthusiasm; the Italians made us very welcome and due to the hot weather kept us well supplied with cold drinks. The older members of the population remembered the B17 in quite a different vein, but nevertheless were very interested in talking about the plane and were delighted to have a chance to have a real close look inside as well as outside. Right up to the last moment we had crowds of very excited Italians still wanting a final glimpse of our famous aircraft. We were all honoured and it was a great privilege to have been made so welcome; a good trip all round.

Air show at Forli, Italy, 1996; Peter is keeping out of the sun

The Big Week

Another very special flight was to Poland to commemorate the uprising in 1944 against the Germans, honouring the courage, determination and sacrifice of the fine people of Warsaw. It began 'The Big Week' as we called it. This is why ...

On the morning of 27th July 2006 we were off to RAF Marham for the Air Force 'At Home Day'. In the afternoon off we went to Lowestoft for the seafront show and back to Norwich for fuel and a night stop. The next day we returned to Lowestoft for the second day of their show, and on completion of our display we set sail for Edinburgh after a night stop. The following day we put on a display at East Fortune, near Edinburgh, and then on to Sunderland for their seafront display; all still going fine. Our next landing was back to Norwich for more fuel. By this time it was our slot into Middle Wallop where an evening concert was taking place; we arrived overhead two minutes late. Fortunately the whole show was running two minutes late as well, so after such a busy day that was not bad.

On the way back to Duxford, No 4 engine started to leak oil - not badly, but with the long flight to Poland early on Monday morning and given that this was now late Saturday evening, my Sunday off wasn't looking too good. So we landed, put the plane to bed and went home for a night's rest.

On Sunday it was back up to Duxford to fix the oil leak and top the oil up. We had enough fuel on board from the Norwich refuel, so I then did a thorough inspection of the complete aircraft and loaded all the spares I thought we might need for the trip. You could never be sure if you had taken enough, too little or the right spares you could need. So with all that completed, I went back home and re-packed my bag for the next flight, which was for Poland.

With everything in order and 1300 US gallons of fuel in the tanks, we started engines at 10.00 hours and checked all the engines and systems: all good so off we went. The planned route was over the North Sea to Amsterdam and Berlin, avoiding controlled airspace.

The weather was good en route and we landed at Okecie International Airport a few miles outside Warsaw. The flight time

FROM PLOUGH TO PLANE

was 5 hours 25 minutes, making us a bit early, but there we were directed to the military apron, as we were to be hosted by the 36th Special Airlift Unit of the Polish Air Force. They are the equivalent of our own 32nd (Royal) Squadron, and their primary duty is to transport Polish Government VIPs.

Warsaw Air Show leaflet 62nd anniversary of Warsaw Uprising

MY YEARS WITH THE 'SALLY B'

The VIPs, Veterans and Sally B crew at the Warsaw Air Show, August 2006; Peter is third from left on the second row

We were greeted by a large crowd of high-ranking members of the Government and Air Force, as well as the media, who made us very welcome. They all wanted to talk about the B17 and what our plans were for the following day. The reason for our great enthusiastic welcome was all that the aircraft stood for, and the fact that Sally B was the first B17 to land in Poland since the Second World War.

View of Sally B dropping leaflets over the city of Warsaw, August 2006

After the reception at the airport we left the Sally B in the safe hands of the Special Guard. So it was off to the hotel for a wash and a good meal in a very good restaurant. Yes, we were really given the VIP treatment. At last we could get to bed after a long day but a very rewarding one, as it is always a great feeling when it all goes well; and it did. As we left the airport there were people streaming in to catch a glimpse of the B17. Some had never seen one, and for the older people of Poland who had, it meant so much to them.

The next day was what it was all about and the reason we were in Warsaw; so we were up for an early breakfast and off to the airport to start work. We offloaded all the spares we had brought over and loaded up thousands of leaflets to drop over the city of Warsaw.

We were going to drop them from a hatch in the floor of the plane. I had worked out how many leaflets we could drop in the time we had, so with all the packets of leaflets lined up by the hatch, we were ready for take-off.

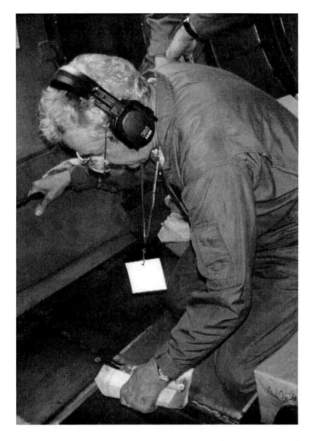

Peter in Sally B dropping leaflets over the city of Warsaw, August 2006

After briefing and a final pre-flight inspection of the plane, we started engines and taxied to line up on Okecie's runway 29 ready to take off for the leaflet drop over the city. The plan was to turn left after take-off and proceed to the south of the airport, where we would make a further left turn to fly along the east bank of the River Vistula, and then at a pre-determined point start dropping the leaflets. I had calculated beforehand that if we dropped a half-bundle of leaflets every fifteen seconds, we would have dropped all the leaflets in the planned flight time; it worked well. I could see the vast crowds looking up at us and the reports from the ground were that there was not a dry eye in the whole of the city of Warsaw. It was a very emotional time for all of us, and the crew of Sally B felt proud and honoured to have been part of it.

Following the leaflet drop we did two further flights for the VIPs and the media. One of the VIPs was the Commander of the Polish Air Force, Lieutenant-General Stanislaw Targosz; after the flight he presented me with a plaque which I was pleased to receive as a great reminder of our visit to Poland. After the last flight of the day we prepared the plane for the return flight the next day, refuelled, topped up the oil tanks, reloaded the spares and checked the aircraft over in general - as we say 'put her to bed'. We were now ready for a good meal, a beer and our bed.

Peter with the Commander of the Polish Air Force, Lieutenant General Stanislaw Targosz, August 2006; Peter is being presented with a PAF plaque commemorating the 62nd anniversary of the Warsaw Uprising of 1944

The next morning it was up early, a good breakfast, and up to the airport to say our farewells to all the new friends we had made in our short stay in Warsaw. We distributed the presents we had brought with us to all the people who had helped us during our stay. That all complete, we started engines and taxied out, waving the Polish flag from the plane as we went. It had been a very successful operation; the plane behaved, the weather was kind, the Polish people were great - what more could we have asked for?

The flight home went without a hitch, although the weather could have been a little kinder. The highlight for me, and I am sure for the rest of the crew, was to fly right over the top of Berlin. Imagine flying over Berlin in a Flying Fortress in 2006! Not many crews have done that. For me, a particular thrill was to fly right over the old Berlin Tempelhof Airport where I had worked on DC4 aircraft for Air Charter, a Freddie Laker airline, during the late 1950s, when the Berlin corridor was still in place. Finally we landed back at Duxford, tired but very satisfied with the week's work.

Down the Mall and over the Palace

Over the years we have done many air show fly-bys, dedicated to many fine people, some of whom have given their all. Each year we fly over the American war graves at Madingley near Cambridge to salute some 79,000 US servicemen who gave their lives during the Second World War.

The really special flight is the one to commemorate the victory of the ending of the Second World War. I have been on the only two that have taken place: first the fiftieth, and in 2005 the sixtieth. It is a very special event to take part in. On the sixtieth anniversary we had the BBC News film crew on board, with Dan Snow; so you can see it is a great responsibility. Once we pass over the Palace I am quite relieved that it has all gone without a hitch. With the world watching, the last thing all the crew and team need is for some technical defect to mean that we have to drop out of formation and not continue with the fly-past. After it is all over and we are on the way back home, it is a great feeling to have played a major part in such a special occasion. When you fly up the Mall and over the Palace and see the many thousands of people down there, it comes home to you what a great event you are involved in, and how lucky we are to fly a great piece of history on such a historical day. My granddaughter was standing by the Palace gate watching the fly-past and that was rather nice.

FROM PLOUGH TO PLANE

Sally B on the way to Buckingham Palace to take part in the VE Day sixtieth Anniversary, 10th June 2005;
Peter was the flight engineer on the flight

View of the B17 fly-past over The Mall, VE Day sixtieth Anniversary, 10th June 2005

FROM PLOUGH TO PLANE

Sally B flying over Buckingham Palace,
VE Day sixtieth Anniversary, 10th June 2005

Manna

Another operation we carried out with our aircraft was to commemorate the end of the war for the Dutch people and the Operation called 'Manna' (the Americans called it 'Chow Hound'). By the end of the war, the people of The Netherlands were starving. The RAF and the US Air Force carried out food drops even though the war was not yet over, something the people of The Netherlands have never forgotten, as it must have saved many lives and helped to relieve some of the hardships they had endured over the years of German occupation.

We went over to help them celebrate, loading the plane up with black plastic bags full of bread rolls, flowers coloured red, white and blue, plus packets of chocolate on small parachutes. Each town and village had paid for the bags to be dropped in their location and each bag was numbered. Some had just one bag, while bigger towns had several bags, so as you can guess, it was not easy to drop them all in the right place. However, we did our very best, as we all knew how important it was for the fine people of The Netherlands.

In addition, we dropped copies of the newspaper that was dropped by the Allied Air Forces on that same day sixty years ago, when the war ended for The Netherlands, with the headlines 'DUITSCHLAND CAPITULEERT', dated 10th May 1945. Once again, it was a great honour to do this.

Reproduction of the last edition of 'The Flying Dutchman', May 1945, bearing news of the capitulation of Germany

The Lunar Rock

From time to time, all of us B17 Flying Fortress operators get together at meetings of the B17 Co-op. This organisation was formed so that we could meet up, discuss the problems we all have, and share information. Another benefit is that if any of us require spares, we can make a bulk order for the whole group, and that way purchasing is more cost-effective.

During one of these B17 Co-op meetings in the USA, I made the acquaintance of a member of the Commemorative Air Force, formerly known as the 'Confederate Air Force', but due to political correctness they had to change their name. (On the other hand, the 'Yankee Air Force', another B17 group, was allowed to keep its name.)

This CAF member was Dr Everett Gibson PhD. As well as being a B17 enthusiast, his main profession is as a NASA Senior Scientist, an expert in Astrobiology. I met him again in 2008 at the Duxford Imperial War Museum, and as we were talking over old times he said, "I have something to show you, Peter." With that he pulled out of his case a leather pouch and inside was a lunar rock sample, which was rather exciting to see. I then asked him if he would take my photograph with the rock. This he did, and so you have this photograph - just a little piece of history. It's not every day that you get to hold a piece of the moon!

Peter holding a piece of lunar rock

MY YEARS WITH THE 'SALLY B'

VIPs Along the Way

Let me start by saying that I don't like 'name dropping' but I must mention some famous people that I've had the privilege to meet. Of course I have met many people throughout my life, some being very interesting 'ordinary' folk (if you could call them ordinary - some were certainly not). I have also encountered a few VIPs. I suppose the first of these was the 'Father' of the Royal Air Force, Lord Trenchard, who was taking the salute at RAF Upwood when I was on parade along with the rest of the 214 FMS Squadron. It was raining, so the parade was held in one of the hangars. When the presentation of arms was complete, he told us to break ranks and gather round him, and he gave a speech to us, just like a father. I had never seen that done before, or indeed any time since then during the time I was in the RAF. For some reason that stuck in my mind, maybe because I was a only a LAC (leading aircraftman) and he was the TOP MAN.

VIP visitor Lord Trenchard, after a flight in a DH86B of 24 Squadron, late 1930s

Other VIPs I have met were mostly as a result of my involvement with the B17. Some of them I met in person, while others were present at displays that we put on for them. They include: HRM the Queen and Prince Philip; President George Bush (Senior); Prince Andrew; and King Hussein of Jordan

During a special air show at Duxford, the crew of the B17 Sally B were introduced to HRH the Prince of Wales. That was the same show that George Bush (Senior) attended, but he left shortly after we had finished our display and were waiting for the Prince, so we never got to meet him.

(Left to right) Roger Mills, Jim Jewell, Peter Brown and Mike Stapley are presented to HRH The Prince of Wales at the Rededication of The American Air Museum on Friday 27 September 2002
Credit: IWM Duxford no DxD2002-08-26

Sally B crew being presented to Prince Charles at Duxford, September 2002; Peter is second from left

I met the Duke of Kent during a Duxford Air Show and I met the Duke of Gloucester at the Fairford Air Tattoo when I presented the crew of the Sally B to him. However, as the Vulcan bomber was doing a display at the same time, the noise was such that I could scarcely hear him, and he couldn't hear me, but I met him just the same.

One of the most interesting and nice men I have ever met was Walter Cronkite, who was the 'anchor man' for CBS news for many years. He reported on the Second World War from England in programmes for the USA, and he would sign off with: "And that's the way it is." I met him at Duxford during the making of a TV programme about the Second World War, and he was as bright as they come.

Peter talking to Walter Cronkite at Duxford, August 2005

Walter was hailed as the most trusted man in America during his time with CBS television as their anchor man. Even when I met him at the age of 88, he was still enjoying the limelight. He covered the John F Kennedy assassination and the moon landings (his favourite story). He went on raids over Germany in a B17 out of Norfolk and he also went ashore with the troops during the D-Day landings. So you can see that he was a great man to talk to and I was very pleased to have had the opportunity to do that.

During an air show we were doing in Czech Republic, the President of Czech Republic at that time, Mr Václav Havel, came to the aircraft to meet the crew. (He is the former anti-communist dissident who swapped his prison cell for the presidency within months of the 1989 Velvet Revolution that ended communism in Czech Republic.) Our air show was in 1994. Mr Václav Havel has since left the Government and gone back into the Arts; he is a dramatist and at the time of writing (May 2008) has a play on at the Prague Theatre called 'Leaving'.

Meeting Václav Havel, 1994; the President of Czech Republic is second from right and Peter is second from left

MY YEARS WITH THE 'SALLY B'

During the making of the film 'Closing The Ring', I had the pleasure of meeting Sir Richard Attenborough. At one stage we thought that the B17 Preservation was going to feature in the film. I was involved in getting the props together for the crash scene (the film was all about a B17 crashing in Northern Ireland) but in the end that was all we did. Nevertheless I did meet Sir Richard and had a photograph taken with him, although, sad to say, the photographs never came out (high-tech cameras!).

Those are just some of the people I have had the honour of meeting, but many more great and brave people have crossed my path over the years. There are far too many to mention; they are the kind of folk who may not be famous, but they are still very special people.

*****

These stories have covered a fair amount of my activities in the B17 Preservation, and I hope they provide some explanation as to why I have been involved with Sally B all these years.

FROM PLOUGH TO PLANE

The Sally B on a memorial flight for USAAF, 2002

MY YEARS WITH THE 'SALLY B'

The 5000th B17 built since the attack on Pearl Harbor, May 1944, Boeing plant, Seattle

Annual Flyover at Madingley War Cemetery, 2000

With an old Flying Machine and its pilot, Duxford, 2003

My Part in the Making of 'Memphis Belle'

My Part in the Making of 'Memphis Belle'

We, that is the B17 Preservation, had heard for some time that a film was about to be made involving the B17 Flying Fortress - a true wartime story. The producer was David Puttnam, but with maybe as many as ten airworthy B17s in the USA, it was very unlikely that we would be involved, given that we had only one flying in the UK. However, as it worked out, the film company could not agree with the American operators, so they decided to make the film in the UK with our B17, two from France and another two from the USA.

The next move was a meeting at Duxford airfield with the associate producer, Eric Rattray, the design director Stuart Craig, Elly Sallingboe, the operator of the B17 Sally B, and myself (Sally B is the name of our B17 aircraft). We are based at Duxford Imperial War Museum. Being the chief engineer on the aircraft, I was to become very involved with the making of the film. It was during that first meeting that I was asked if I would take on the role of aeronautical consultant for the film. I was more than pleased to say yes as it would involve going to the States to inspect the two aircraft that would be used in the film. The reason for this was to ensure that both aircraft were up to the legal and maintenance standards required by both the film company and the UK Civil Aviation Authority.

The two aircraft from France I had seen over the years at air shows and I knew the engineers that maintained them, so they did not cause me any concern. During this time I was working full-time for HeavyLift Cargo Airlines at Stansted as the line maintenance manager, so it was going to be a busy time ahead.

FROM PLOUGH TO PLANE

The Sally B

MY PART IN THE MAKING OF 'MEMPHIS BELLE'

The Sally B B17G crew and team; Peter is 3rd from left on the back row

The film at first was going to be called 'Southern Belle' but it was later changed to 'Memphis Belle'. Memphis Belle was the first aircraft to fly 25 missions during the Second World War with the USAAF, from Bassingborne in Cambridgeshire. The producer for the film was Catherine Wylder, daughter of William Wylder; he made a documentary about the aircraft during the Second World War. The name of the aircraft was the girlfriend of the captain - his name was Bob Morgan.

So now after a few more meetings we were ready to start getting the show on the road. We still had the air show season to contend with; however we could still carry out some of the modifications required for the film. That list was long but we did all that the film people had asked for; all went well.

Now that our aircraft was all under control, I packed my bags and headed for the USA. The first plane was at Chino, California, a very interesting place with many warbird planes and a great museum. However, my task was to carry out the inspection on that one B17 to make sure it was fit to fly over to England and carry out the fifty or so flying sessions we required it to do for the film. It wasn't in the greatest of maintenance standard, so on that visit I said I would have to carry out another audit when the required maintenance had been carried out.

The next port of call was to Geneseo in New York State. The owner of this aircraft I knew well; he was a good friend of mine. That aircraft was in a great state of serviceability but by this time Austin Wadsworth, the owner, had decided that he couldn't let the aircraft come over to the UK, with his busy air show season about to start.

What now? Austin called another B17 operator in Seattle. Maybe he would like to bring his plane over to England? At first he wasn't all that keen, but he was interested. I couldn't say yes or no as I had to check it out with the film boss and I had to get back to Stansted for my real job. I had left Stansted at 2 o'clock on the Thursday and was back in my office at 10 o'clock on the Monday - must be a record.

The next task was to get Sally B ready. We finished the winter maintenance, completed the work required by the film company, did engine ground checks, the air test, and then we were ready for the big day.

MY PART IN THE MAKING OF 'MEMPHIS BELLE'

Working on the 'Memphis Belle'

I went back to the States and re-inspected the Chino plane, and this time I went to Seattle. That aircraft was in very good condition, so that was no problem. With the Chino plane up to the satisfactory standard, we were ready to go.

Having got the two aircraft in the USA sorted out, I was now able to spend more time on our aircraft. At that time the aircraft was in the wrong colour scheme and had to be repainted. We couldn't do this at Duxford so the aircraft had to be flown to Southend. Not only did Sally B require a repaint, but so did the two French B17s, which meant that at this time we had the two French planes at the paint shop as well as ours.

As I said before, I was still working full-time as chief engineer for HCA, and they had their main maintenance base at Southend. On the first morning of filming at Duxford we still had the two French B17s at Southend. Now I had to somehow get down to Southend to sign out one of the company CL44s, a large cargo aircraft. I did intend to drive down, but after telling the producer what I was about to do, he said, "Hold on, why don't we fly Sally B to Southend from Duxford? You can do what is required by your company and then we can fly back to Duxford in formation." So that is what we did - a great start.

The Chino aircraft was already at Duxford, so now we had four; the Seattle aircraft was somewhere between Seattle and Duxford. Later on that first day we heard that it was on its way from Prestwick in Scotland. Everything was falling into place, or so it seemed, but on the last few miles from Duxford, one of the engines failed; they had to shut it down and continue on three. That was no big deal for a four-engine aircraft, but it did mean it would require a new engine before we could start filming.

The only spare engine at Duxford, or come to that anywhere this side of the States, was our spare. After doing the deal for the Seattle aircraft to have our spare, being the engineer looking after the five B17s, I now started to organise the change. With the help of the crew and some of my engineers we changed the engine, carried out ground checks, air tests, and at last we had five serviceable B17s on the ramp ready to start rolling, or as Catherine Wylder said: "HERE WE GO INTO THE WIDE BLUE YONDER."

During the next few days all went fine but then one of the French aircraft also had an engine problem, which resulted in a cowling (a large panel around the engine) falling off and landing in a field near Diss. So once again we were back to four, although another engine change was on the way. Now the only problem was that the film people had used the engine to make plaster casts to mock up props for other shots during filming. Unfortunately some of the plaster had got into the engine oil system and caused the French engineers

MY PART IN THE MAKING OF 'MEMPHIS BELLE'

all sorts of trouble in trying to clear the engine. With the engine now serviceable, and the air tests satisfied, we were back to five, all ready to complete the Duxford side of the filming before the move to Binbrook, Lincolnshire, for the second part. Duxford had been where the flying scenes were shot, and Binbrook was where the ground filming would take place.

So on the Sunday all five took off from Duxford. On our way we did a fly-past over the American war cemetery at Madingley to make a salute to the American crews lost during the war.

On the way up to Binbrook we met the 'Red Arrows' going towards Duxford. They gave us a smoke welcome and as we had fitted a smoke system to two of our engines, that meant we could return the welcome to them.

On arrival at Binbrook we were the first to land, and while we were getting ready for the landing I went back into the bomb bay to fill up the smoke tanks. But when I went back into the flight deck and looked out of the window, I saw four B17s already parked up. I thought they had been quick to get there, but on landing I realised that they were mock-ups. They were so realistic that it was really hard to tell them from the genuine article from the air. So here we all were safely on the ground and ready for the next three weeks' filming.

Binbrook was an ex Royal Air Force station with all the buildings, such as the mess hall, hangars, control tower, a barrack blocks - in fact a complete airfield. Just the place to make a film about a wartime base.

The filming went as planned but as I was still working full-time for the airline I had to go to work for the first week. It was during that week that the French B17 taxied into our plane and broke the glass nose. We had a spare one at Duxford so I arranged for that to be taken to Binbrook, went to check it was fitted correctly and signed for it. Then the next morning I left very early to drive to Southend for an engineering meeting. All in all I was glad when that week was over and I was on leave for the next two weeks - the time it would take to complete the film.

There was an accident during that two weeks when one of the B17s from France crashed on take-off. Luckily no one was killed but several had bad injuries, broken legs etc. It did not hold up the film and shooting went ahead the same day. It was a great shame to see such a great aircraft go up in flames and be totally destroyed. The only part left was the rudder, which we still have as a spare. I hope we never need it.

As I have said before we didn't play the role of Memphis Belle until the last scene, and that was when the aircraft was returning from its last mission, the 25th. Now what they did was to shoot us as though we had just landed all shot up. We had fitted a mock fin and rudder, badly damaged, and had smoke pouring out of the engines. In fact we started at the end of the runway with two engines already stopped and then raced down the runway and stopped the third engine. So with only one engine going we came off the runway onto the grass and came to a halt. Then the next scene was with all the actors getting out of the aircraft after their last mission; and that was the way it was done.

I got a small part in the film, being seen once as an engineer working on the engine. So what with that and with getting a credit as the aeronautical consultant, I was very pleased with it all.

Peter's acting role during the making of 'Memphis Belle'

MY PART IN THE MAKING OF 'MEMPHIS BELLE'

With the B17 crew, winning the cup at the West Malling Air Show. From left: John Littleton, Crew Chief; Mac Kiney, B17 Captain; Mike Collins, Belfast Captain; and Peter, B17 Chief Engineer

HeavyLift Cargo Airlines

HeavyLift Cargo Airlines

During my time working for a cargo airline specialising in outsize loads, we carried an amazing variety of cargoes. In the main, the cargo was for the Ministry of Defence, such as helicopters, tanks, aircraft-towing tugs, fuel bowsers, buses, and even at times bombs.

CL44 G-AXUL being loaded at Stansted, 1979

Other cargo we carried included ballot boxes, when there was a election in a friendly country. At one time we were taking gold out of the country every week for quite a few weeks - twenty tons at a time. It was not, as you would expect, all packed in boxes, but just loaded onto pallets for all to see. However, the gold loads always had some security guards with them, so I guess it would have been hard to run off with a bar or two.

One of two Bell 212 helicopters carried from
Stansted to Brisbane inside the CL-44 Guppy, 1989

Belfast being loaded with Fokker aircraft fuselage,
The Netherlands, 1998

Satellite dish being loaded onto a G-BEPS

Bus being loaded onto Belfast aircraft, Stansted, 1999

FROM PLOUGH TO PLANE

The CL-44 Guppy carried two of these barges, each weighing 9500 kgs, from Europe to the Middle East, and later flew similar loads to Alaska, following the major oil spillage there in 1988.

One of the most interesting loads we carried was a complete circus (and I mean complete - from elephants to clowns). It was a Gerry Cottle circus that the Sultan of Oman had hired for his birthday party. It was too large for one aircraft to take, so he had to hire two. It all came to Stansted to be loaded onto the aircraft, but bad weather delayed the departure for a day or so. It was about what happened during that time that I have a story to tell. At the time we all thought it was funny, but now I don't know.

I was on night shift at the time; the shift consisted of a full team of maintenance personnel, engineers, cleaners, and a storeman. If the engineers required a spare to fix a defect, he would go to the stores and draw out the spare. Every time we went to the stores counter, the hatch was locked and we had to knock, or rather most times hammer on the hatch door to wake the storeman up. He would eventually open up and say he had been out in the back sorting out something or other, but in fact he looked as though he had just woken up.

This night the circus equipment was still in the hangar waiting to be loaded, and among it was a gorilla suit. One of the engineers put the suit on, went to the stores and hammered on the hatch, which was, as usual, locked. Our storeman came out sleepy-eyed and opened the hatch, whereupon the 'gorilla' made an attempt to climb into the stores. You can guess what effect it had on the poor storeman! As I have said, it may not have been the wisest thing to have done thinking about it now, but on the other hand he did keep the hatch open after that.

The next night the weather was good, so all the animals were loaded, plus all the other circus props and, yes, including the ape suit. Off they went to the desert. The Sultan was so satisfied with the circus that he bought the lot and it stayed in Oman. Actually I don't think he kept the clowns, but you never know.

Over the years we carried many types of animals from whales to day-old chicks, pigs, sheep, cows, racehorses, monkeys and many more. One particular flight I shall always remember was when we transported a load of giraffes. They had originally been booked onto a Boeing 747, but by the time the documents for their shipment had been sorted out, they had grown too tall for that aircraft. So they came to HeavyLift Cargo Airlines and we transported them in the CL44 Guppy.

Another contract I went on was to Barranquilla in Colombia, South America. The reason we were there with two Belfasts was to transport oil-drilling equipment from Barranquillia over the Andes to Alrolka. The national airline of Colombia was not very happy with us working in their country. They made an objection to the authorities, saying we were taking work away from them. However, as they didn't have any aircraft that could carry anything the size the oil company wanted us to transport, they didn't have much of a case; but that does happen in the cargo world.

FROM PLOUGH TO PLANE

Giraffes being loaded aboard a CL-44 Guppy, 1987

CL-44 Guppy, 1987; one of the 21 giraffes and 38 antelope that were flown aboard this aircraft from Kenya to New York. Just checking the airline for next time.

HEAVYLIFT CARGO AIRLINES

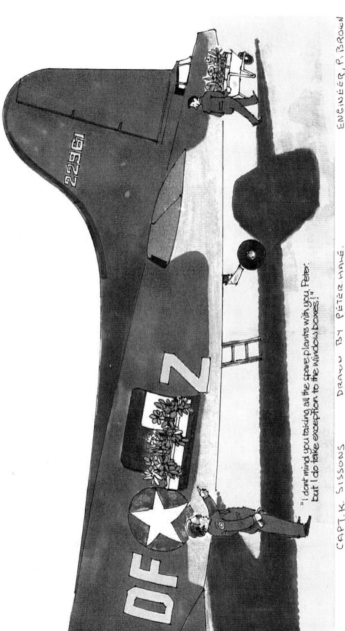

Cartoon of Captain Keith Sissons and Engineer Peter Brown, West Malling Air Show. 1990

FROM PLOUGH TO PLANE

Colombia was a very interesting part of the world to work in, and the people seemed happy with their lot; the engineers I worked with at the airport were very helpful. I was quality control manager at that time and had helped a local company to set up a workshop to overhaul our main wheels. That gave me a good insight into their working procedures. I have found in the airline business that no matter what part of the world you are in, engineers will always help each other out, even if you can't speak the language, which was often the case. I have spent many a day working under such conditions with no great problems.

As soon as the price of oil fell, the oil company started to ship the equipment by road and barge, a much slower method but a lot cheaper. So we packed our equipment spares and the like, said our goodbyes and it was once more back to base.

Another little tale I must tell about cargo is the monkey saga. During my time working at Southend airport I was with a airline named 'British United'. They operated the DC6 aircraft and one of the contracts we had was a freight run down into Africa. Our first stop was Nairobi, Kenya and then it was on to Salisbury (now Harare) in Rhodesia (now Zimbabwe). The load outbound was general freight, but the cargo back to Southend was a live one - hundreds of rhesus monkeys. The aircraft was completely packed out with them in cages.

During the offload, some would get out and make for the roof of the hangar; they then became very hard to catch. One day there was one we couldn't catch no matter what we did, so the lorries that came to collect them had to leave one monkey short, although given there were hundreds of them, one missing wasn't a big deal. One of the engineers managed to catch it later after a day or so, took it home and had it for a pet; he was often seen walking around the village with it on a lead.

However, our problem was that the aircraft was required for a passenger flight a few hours after landing. Once the last monkey was off the aircraft, the cages were removed and the floor swept clean. We then had to lay the carpet, fit the seats, galleys, toilets, and the overhead shelves (no overhead lockers at that time). The next job was to give the passenger cabin a really good spray with some sweet smelling scent - maybe roses or the scent of the day - and the aircraft was ready for its next flight. It was the given a turn-round inspection, refuelled, passengers were loaded on board, the engines were started, and off they went on holiday.

I don't think many would guess that a few hours beforehand, their holiday plane had been full of monkeys. On the other hand, maybe they wouldn't have cared; they were off on holiday, smelling of roses!

During my last years with HeavyLift at Stansted, I spent many weeks away from home carrying out engineering audits around the world at various airlines' engineering bases. The reason for this was that we were using foreign airlines and my task was to ensure that they had equivalent safety standards as required by the European Union.

My role in this was to visit the overseas airlines, carry out an inspection of their engineering base, and make recommendations to them so that when the UK Civil Aviation Authority made their visit to grant the approval, all would be up to the required standard.

Let me tell you about some of the places I visited during that time. My first inspection was at Singapore, the airline being Pelita Air Services of Indonesia. The aircraft was a C130 and its task was oil spill protection for that part of the world. My company had the contract with the oil companies, so it was our responsibility to ensure all was satisfactory.

The Singapore contract also took me to Indonesia as that was where the aircraft was maintained, so that had to be up to the CAA standard. I was up and down to both Singapore and Indonesia every two months or so. I found the engineers out there very pleasant to work with and good engineers. Their system took some getting used to with seniority within the workforce, as that seemed to count for everything.

The base in Singapore was at Seletar, an old Royal Air Force flying boat base. All the roads were named after English places; it was a really lovely place to be stationed.

During the Singapore contract the Pelita aircraft had to be returned to their air force, so there we were under a contract with the oil company and no aircraft. The only civil C130s not in the military service were in South Africa, owned by an airline called Safair, so that was the next port of call.

I flew from London to South Africa to do the required inspections on the aircraft and complete the documentation to make sure once again it was up to the standard required by both the CAA and the oil company. It took a week to do that and then I went with the aircraft to Singapore. I got there the day the other plane left for Jakarta so that worked out fine. On landing we had to prepare the plane for a cargo that night.

After a while HeavyLift lost the oil spill contract to the South Africans, so that was the end of the Singapore flights.

On board Concorde Reg G-BOAC from Heathrow to Sweden 'Vasteras' (Stockholm), September 1987

HEAVYLIFT CARGO AIRLINES

Sally B in formation with Concorde at 'Vasteras' Sweden (Peter was the Flight Engineer on the B17)

The B17G Reg G-BEDF flying over Concorde (Peter was the Flight Engineer on the flight)

HEAVYLIFT CARGO AIRLINES

In Singapore outside the POW chapel in
Changi prison camp

In Barranquilla, Colombia, 1985, when working for Heavylift
Cargo Airlines

FROM PLOUGH TO PLANE

Winners of the West Malling Air Show Bob Richardson Cup, 1991, with HeavyLift's Belfast aircraft. From left: Captain Mike Collins; Flight Officer Pete Heming; Peter, Chief Engineer; Flight Engineer Louie Evans (picture taken at Stansted)

The winning team in colour, Stansted, 1991

The next contract was with the Russians as we (that is, HeavyLift Cargo Airlines) had a joint venture with a Russian airline, Volga-Dnepr. It is named after the two rivers that flow through the town that is their base - Ulyanovsk - which is also the birthplace of the great Russian leader Lenin, or should I say Vladimir Ilyich Ulyanov, his real name. So now you can see why it is called Ulyanovsk, but that's enough of the history lesson for today.

My first visit to Uly, as we called it for short, was just before Christmas. HeavyLift and Volga-Dnepr together were doing the Operation Christmas Child present run, when people in the UK fill hundreds of shoe boxes with Christmas presents for the children in Russia. The aircraft was the large Antonov 124, at that time the largest aircraft flying, so that was a good way for me to get there.

I was not personally involved in the Operation Christmas Child present run. My task was to carry out the engineering audit on the engineering base, which went very well, as the Russians have very well trained engineers. Their systems were not the same as in the UK, but provided it were the equivalent, that was all I wanted.

Being the middle of winter, it was cold. The River Volga was frozen solid and the fishermen had cut holes in the ice to go on fishing. It was so cold in the crew bus - what it was like out there on the ice did not bear thinking about.

On my second visit I went from London to Moscow by British Airways and then by a local airline from an airport on the other side of Moscow - a long ride in the winter and on the not so good Russian roads, but we got there in the end.

The plane was a Yak 40 and when I was boarding, standing at the bottom of the stairs was a man with a large dog, and believe me it really was big. I didn't think too much of it at the time, just that it must have been a guard dog. But as I was getting in my seat I heard this noise, looked around, and yes, the dog was in the seat behind me. It seemed quite happy. I wasn't over the moon, as they didn't put a muzzle on it until we were on the ground at Uly.

It was a rather strange experience travelling on the local airlines. One thing was that the crew always leaves the aircraft first and no one moves until all the crew are off, and that includes the cabin crew. By now they may well be more in line with international standards, but it was all right and I enjoyed most of it.

The vodka drinking was something else; the Russians do enjoy a glass or three. The only way I could manage was to fill my glass with water when no one was looking and carry on as if I was keeping up. They must have thought I could take my vodka. They were really good people and it was great to have worked with them.

The joint venture with HeavyLift lasted ten years, so you can see I got to know Uly and the Russians very well. But nothing lasts for ever; in ten years the Russians got very good at the cargo business and went on their own, which meant I had to say goodbye to them all. So what next??

HEAVYLIFT CARGO AIRLINES

A *dacha* in Ulyanovsk, Russia (in a KGB holiday resort), where Peter stayed on one of his trips

Once that joint venture with the Russian company was over, HeavyLift Cargo Airlines had to look for another partner to continue in the outsize market. The only company operating the large AN124 aircraft other than the military was the Ukraine company Antonov Airlines. They not only operated the aircraft but they were also the manufacturers. They also operated the giant AN225, which could carry some 225 tonnes of cargo. When I was flying on the DC4 out of Southend to Australia, our load was about 12 tonnes.

So my next audit was in Kiev to carry out a pre CAA-audit. The Ukraine's system was much the same as the Russian one as far as the ground engineers' training and maintenance procedures went, so that made it easy as far as that was concerned. After about four days I thought we were ready for the Civil Aviation Authority of the UK.

Peter having his collar felt by Captain Webb, Russia, 1995

Outside Lenin's house in Ulyanovsk, Russia, 1995

FROM PLOUGH TO PLANE

In front of Lenin's statue as a child with his mother, Russia, 1995

HEAVYLIFT CARGO AIRLINES

The next step was to go back to Stansted and wait for the CAA. Within a week or so I was back with the CAA team so they could carry out their inspection and satisfy themselves that Antonov Airlines were up to the required standard to comply with the EEC directive. There were a few days of inspecting and sightseeing - more sightseeing and less inspecting, that's sometimes the way it goes - but at the end of the day a good inspection had been carried out and everyone was happy. The airline got the documents stamped, the standard was maintained and we all went home.

My last audit was in Egypt looking at a company named Cairo Air. They operated a Russian TU204 with Rolls-Royce engines.

I had been to Egypt a few times since my Air Force days but that had been to fix defects on one of our aircraft. In the aircraft world as soon as the aircraft is serviceable it is back in the air earning money - no time for resting or sightseeing. In my time as an aircraft engineer I have been all over the world to fix aircraft and many, many times I have only seen the airport as you arrive with your toolbox and the spares you require to fix the defect under your arm. The operating crew will be waiting to take you straight to their aircraft; you fix it, hop on board and off we go back home.

But with the audit side you don't have planes to repair. It is a bit more relaxing and you have time for a look around, and a bit of shopping. When I was in Cairo I didn't manage to visit my old station at Shallufa. We said we would go next time but then there wasn't a next time. However, they do say, "never go back," and I think that's true, because when I went back to my old station at RAF Upwood, it was a bit disappointing. It was not the same as when I was a young airman.

* * * * *

As a postscript - one more thing about Egypt and the Canal Zone episode that I should add is that after some fifty years, I quite recently received the General Service medal with the Canal Zone bar. Truly the wheels of the Government and the Ministry of Defence do turn very slow!

Peter returns to the old Guard Room, RAF Upwood, after fifty years: "Not the same - the old Flight Sergeant would go mad at the state of it!"

The Changing Village

The Changing Village

Village life has changed so much since I was a lad. Every time I return to the village of my birth, I still can't believe how much. In fact the farming way of life bears no relationship to when I was working on the farm for those few years, first on a small farm and then on a very large one.

The biggest change I notice is that you don't see anyone around the village, and no one is actually working in the fields. You used to see teams of men working in the fields doing such jobs as hoeing the young plants of sugar beet. Now, when the sugar beets are ready to harvest, a plough-type implement lifts them out of the ground, shakes all the dirt off them, cuts the tops off, and then the beet go straight into a trailer alongside the sugar beet harvester; two men are all that is needed to complete this task, instead of the ten or so needed when I was a lad working on the farm.

Before mechanisation, the men would pull the beets out of the ground by hand two at a time, knock them together to remove the earth from them, and lay them in lines. Next, using a special type of small sickle with a hook at the end of the blade, they would pick the beet up with the hook, then cut off the leaves (or as we would say, the 'tops') and put these in heaps. The tops would be carted off the field to feed the cattle, a job that was done by the likes of me, the boy of the team. You would load the tumbrel up and take it to the field where the cattle were. Now that was fine, but as it was usually the last job of the day, the horse thought it had finished for the day. If you had to pass the farm on your way to give the cattle their feed, it was one hell of a job to get the horse to pass the farm gate.

Finally, the men would pile all the beets up in large heaps ready for the lorry to take them to the sugar beet factory in Bury

St Edmunds. So you can see that it was an operation requiring a large amount of labour; and what a backbreaking, cold job it was! They would wrap a sack around the lower part of their body just to keep the cold out, and when it rained, the sacks got wet. Now two men with the modern machines can do it all in comfort.

Haymaking was another labour-intensive task that not so many years ago required several farm workers. The grass was cut by a horse-drawn cutter and then turned over to allow the sun to dry it out. It was then made into haycocks by men with pitch forks (long-handled and two-tined). After standing in the fields for a few days, the hay was ready to be carted. It was then loaded onto wagons and taken to the stack yard, where it was made into hay ricks. The hay ricks were then thatched to keep out the weather. When you needed hay to feed the cattle, instead of opening the whole top of the rick, we used a large hay knife to cut a section of about a yard square, and that way the rest of the rick was still protected from the weather. At one time my job was to look after a yard of bullocks, and this is when I leant the art of using the hay knife.

But I suppose the biggest job was at harvest time; so let me explain how much corn growing and harvesting has changed. Once the land had been ploughed, it was ready for sowing, a job which hasn't changed too much, except that the sowing drills are larger and more modern. However, that is where the similarity between the past and the present ends as far as the growing and the harvesting of the corn crop goes.

Before the days of weed killer, large gangs of men and women would walk up and down the fields pulling up all the weeds. For the docks they would have a docking fork, a small two-pronged fork (docks, having such long roots, are too hard to pull out by hand alone).

Hedging and ditching was a job carried out when all the crops had been harvested and work was a bit slack, not that it wasn't a very essential task - on the contrary it was very much so. The 'ditching' was when all the ditches were cleaned out, so that all the rainwater running off the fields could find its way to the rivers. Most of the water flowing off the fields ran through clay land-drains; they were about a foot long and were laid underground. The other system was mole draining; for that operation a plough-type of machine was used. It had a knife-shaped tine with a bullet-shaped piece of metal fixed to its end that was dragged through the earth about two feet down. This made a tunnel for the water to drain into, and from that into the ditch, so you can see why it was called a mole.

At the same time as the ditching, the hedging was also being carried out; now that was a real craft, as the farm worker could cut and lay a hedge so that on completion it looked really tidy. It was good for the growth of the hedge, unlike today, when a cutter on the back of a tractor rips and tears the hedge to pieces and they look like some giant rats have been gnawing at them. But I suppose we have to have progress (if you can call it that).

One other thing - with all the heavy machinery on the fields these days, I wouldn't think many clay drains are still in working order, and what with the ditches not being cleaned out, it's no wonder that there is a lot more flooding than used to be the case.

It is in the cutting and harvesting of the cornfields that things have really changed dramatically. Whereas it used to take twenty men and three or four boys, just two men can do the lot today. To start off the cutting, a man with a scythe would cut a strip of land all around the cornfield, so that the horse pulling the binder would not trample down the corn. Once that was done and the cut corn was tied up, the rest of the field was ready for cutting. The binder would cut the corn, tie it up into sheaves, and throw them out onto the ground. Then the men would pick them up, two at a time, and stand them with the ears uppermost into shocks consisting of about, say, twelve sheaves. There they would stand for a few days just to fully ripen before being carted off to the stack yard.

By this time (early 1940s), some of the farm wagons had their shafts removed and a towbar fitted, so instead of having a horse pulling them, a tractor could take its place. The only drawback here was that whereas a horse would go from shock to shock guided by a word from the loaders, a tractor had to have a man driving it - perhaps one of the only jobs where there was a increase in manpower. To load the wagon, two men would pass the sheaves to another man in the wagon, getting as many sheaves on board as the horse could pull. With the tractor, you could load up as many as you could get on board; the only limitation was you couldn't go too high, as the load would topple over, which clearly wasn't good. The other thing to watch out for was to make sure that when the wagon crossed a furrow, both wheels crossed together. If not, the wagon would tip to one side and then the load could move.

The loaded wagon would then take the sheaves to be stacked in the farmyard, but if this was too far away from the field being harvested, sometimes the stack would be built in the corner of the field. The first thing you did was to make a base of clean straw so that the sheaves would be off the ground, to prevent them getting wet. After this the wagons would be offloaded and two men would start to build the stack. When the stack had reached the correct height, the stack men would finish by building a roof for the stack similar to a house, and then it would be thatched - a real work of art. Farm labourers were always classed as unskilled but not many people could build a stack as upright and finished as they did.

There was always lots of competition between the farms on the standard of the stack building. If by chance any farm had a stack that required a prop to prevent it falling down, the stack builder and the farm never lived it down. That stack was always the first to be threshed to get the embarrassment out of the way. That doesn't happen in today's farming; only square or round bales covered in black plastic are to be seen around the countryside.

Once the field was all clear of the shocks, a horse-drawn rake would then rake all the loose corn up, and that would be taken to the stack. The villagers were allowed to glean whatever was left, which was not a lot, but as most people kept chickens, it would feed them for a while. So now all was safely gathered in and all that was left to complete the cycle was to get the threshing done.

Not all farms owned a threshing machine. Down our way, the threshing was done by a co-op and they would go from farm to farm doing the threshing as the farmers wanted.

The 'threshing tackle', as it was known, consisted of a steam engine to pull it along and to drive the threshing drum and elevator. This machinery passed the threshed straw to the men at the rear of the drum, who built a stack of straw, which was used to feed and bed down the cattle during the winter months. (At that time we got some really hard winters and the cattle were kept in the big yards close to the farms.) Once the threshing tackle was in place, with the drum alongside the stack, and the elevator in place where the straw was to be discharged, the steam engine was coupled up. This was done via a big belt from the engine to the drum dead in line with each other - if not the belt would not stay on. If the belt came off it could do great harm to anyone nearby; we hadn't heard of Health and Safety at that time.

Now we were ready to start. Two men on the stack would pass the sheaves to the man standing on the drum. His job was to cut the string holding the sheaves together and feed it into the drum, where it would be threshed. The corn would come out of a chute at the front of the drum, the straw at the rear and the chaff at the side. The chaff man was nearly always the boy - it was a dirty old job. I did it a few times and the worst time was during the threshing of barley, with the havells getting everywhere (the havells are the needle-like spikes on the ears of the barley, and would be very uncomfortable wherever they got to).

During the time the stack had been standing, rats and mice would have been making their way to the stack, where it was nice and warm for them, with plenty of food. But they were not that welcome as far as the farmer was concerned, so before the threshing started, the farmer would put a wire netting fence all around the complete operation. This ensured that when the rodents ran out of the stack, they could get no further than the fence, where either the farmer's dog would be waiting, or the village boys with big sticks. After all they were vermin and did a great deal of damage.

Now you can see why so many people were required to gather in the harvest and carry out many other tasks during the farming calendar. In contrast, just one man in his present day air-conditioned combine harvester can do the work of some ten or more men in comfort.

There is one more thing to add about farming. During the war, and when food was still on ration, when we were doing any of the major tasks, such as at harvest time, the Government would allow us extra rations. So they must have recognised the hard work the farm worker did.

FROM PLOUGH TO PLANE

Threshing tackle machine working at Hillside Farm, Market Weston, 1939

THE CHANGING VILLAGE

Cutting the cornfield at Hillside Farm, Market Weston, 1940

FROM PLOUGH TO PLANE

Harvesting today in 2008 with a modern combine harvester - just one man and his machine

The One Place Where It All Comes Together

The One Place Where It All Comes Together

Now what could I mean by that? Let me explain. I was born in Hopton but spent most of my childhood and up to the age of 17 living in Market Weston. During that time I worked on a small farm, Hillside Farm in Market Weston. For a number of years I had been heavily involved with the Boeing B17 Flying Fortress and the 388th Bomb Group based at Knettishall, Suffolk.

My friend, the farmer David Sarson, who farms Hillside, has over the years built up a very interesting museum dedicated to the 388th BG. That museum brings my life together in some small way - the village, the place where I worked as a young lad, the wartime involvement with the Americans, the 388th BG - right up to the present day. I am still the Chief Engineer of a B17 (the only airworthy one in the UK) and because of that I have been able to help out with finding and donating B17 artefacts for David's museum, which makes me part of it, and proud to be so.

The museum is housed in two ex-wartime Nissan huts (or 'Quonset' buildings as the Americans called them) obtained by Mr Sarson, David's father, after the war, when Knettishall was being returned to the farming world. Thus the buildings and all the contents are genuine wartime artefacts and connected to Market Weston and the 388th Bomb Group.

It's an excellent small museum, which the veterans from Knettishall still visit on a regular basis after some 65 years, so the involvement is still there with the local population. That is what I meant by the 'one place', and I hope I have explained why this is the case.

At Hillside Farm Museum:
from left, Peter, Vera and David Sarson

Ken in thought:
Hillside Museum, 1999

Summing It All Up

Summing It All Up

The year is now 2008, and the reason for this last page is to bring you up to date about the book, now that I hope you have read it and enjoyed my story.

The first thing I would like to do is to thank my wife for her support throughout our married life, and putting up with my being away from home; also for our much-loved family, Paul, Teresa, Jacqueline and Valerie.

I have enjoyed my life from my early days with our big family, my school days at Market Weston until I was 11, and then Barningham until I was 14. Most of that time was during the Second World War, but even so, this limited education gave me a start in life for which I was always thankful.

My few years working on the farm was an education in itself and I suppose I grew up a bit during this period.

The five years in the Royal Air Force was a great experience for a lad who had never been very far outside East Anglia; but the biggest plus was that I was put on the road to a career in the civil aircraft maintenance field.

My working life as an aircraft engineer in the early days was quite hard, with long hours, but very enjoyable. I gained my aircraft maintenance licence in 1967 on a Douglas DC4, and over the years added large aircraft such as the Boeing 707, Douglas DC8, Short's Belfast, the Canadair CL44, and many more.

After I retired from the airline world, I continued working on light aircraft and obtained a wooden fabric licence on planes such as the Tiger Moth and all that type of aircraft.

I am still the chief engineer on the only flying B17 in the UK, so with that I am still very much involved in the aviation world.

So that, I hope, will have brought you up to date.

Peter Ellis Brown LAME
(Licensed Aircraft Maintenance Engineer)
4047596 SAC RAF Retired
Africa General Service Medal: Kenya Bar
General Service Medal (1918-62): Canal Zone Bar